Handbook of Applied and Abstract Analysis

Volume I

Handbook of Applied and Abstract Analysis
Volume I

Edited by **Charlin Chester**

NY RESEARCH PRESS

New York

Published by NY Research Press,
23 West, 55th Street, Suite 816,
New York, NY 10019, USA
www.nyresearchpress.com

Handbook of Applied and Abstract Analysis: Volume I
Edited by Charlin Chester

International Standard Book Number: 978-1-63238-227-6 (Hardback)

Printed in China.

Contents

Preface

This book is a mathematical text reviewing the traditional form of analysis. This book comprises the latest researches and solutions recommended by visionaries across borders in this particular discipline. The book gives equal significance to both; the theoretical discoveries and their applications in the modern world in various fields of science and technology and engineering; to name a few.

The book discusses in detail various forms of analysis such as linear and nonlinear, both ordinary and partial differential equations, optimization theory, and control theory. The relation between optimum control of pseudoparabolic variational inequalities along with asymptotic behavior of higher-order quasilinear neutral differential, Q-symmetry and conditional Q-symmetries for boussinesq equation are significant methods of evaluation. The molecular adiabatic compressibility of methyl and ethyl caprate under high pressure and temperature are depicted under macroscopic expressions. The book is a complete enclosure of all existential theories of mathematics. Some central applications of abstract and applied analysis are iteration methods for time-fractional modified nonlinear kawahara equation and fractional differential equations are important for conducting a theoretical and applied examination.

I convey my regards to all the authors and editors of this book who made the effort of contributing their time and knowledge for this book. I also wish to thank the publishing team for their excellent technical assistance as and when required. Lastly, I wish to thank my family for their support.

Editor

1

Asymptotic Methods for Solitary Solutions and Compactons

Ji-Huan He

National Engineering Laboratory for Modern Silk, College of Textile and Engineering, Soochow University, 199 Renai Road, Suzhou 215123, China

Correspondence should be addressed to Ji-Huan He, hejihuan@suda.edu.cn

Academic Editor: Dumitru Băleanu

This paper is an elementary introduction to some new asymptotic methods for the search for the solitary solutions of nonlinear differential equations, nonlinear differential-difference equations, and nonlinear fractional differential equations. Particular attention is paid throughout the paper to giving an intuitive grasp for the variational approach, the Hamiltonian approach, the variational iteration method, the homotopy perturbation method, the parameter-expansion method, the Yang-Laplace transform, the Yang-Fourier transform, and ancient Chinese mathematics. Hamilton principle and variational principles are also emphasized. The reviewed asymptotic methods are easy to be followed for various applications. Some ideas on this paper are first appeared.

1. Introduction

Soliton was first discovered in 1834 by Russell [1], who observed that a canal boat stopping suddenly gave rise to a solitary wave which traveled down the canal for several miles, without breaking up or losing strength. Russell named this phenomenon the "soliton."

In a highly informative as well as entertaining article [1], Russell gave an engaging historical account of the important scientific observation:

> *I was observing the motion of a boat which was rapidly drawn along a narrow channel by a pair of horses, when the boat suddenly stopped—not so the mass of water in the channel which it had put in motion; it accumulated round the prow of the vessel in a state of violent agitation, then suddenly leaving it behind, rolled forward with great velocity, assuming the form of a large solitary elevation, a rounded, smooth and well-defined heap of water, which continued its course along the channel apparently without change of form or diminution of speed. I followed it on horseback, and overtook it still rolling on at a rate*

of some eight or nine miles an hour, preserving its original figure some thirty feet long and a foot to a foot and a half in height. Its height gradually diminished, and after a chase of one or two miles I lost it in the windings of the channel. Such, in the month of August 1834, was my first chance interview with that singular and beautiful phenomenon which I have called the Wave of Translation.

His ideas did not earn attention until 1965 when Zabusky and Kruskal [2] began to use a finite difference approach to the study of KdV equation, which was obtained by Korteweg and de Vires [3]. Various analytical methods also led to a complete understanding of solitons, especially the inverse scattering transform proposed by Gardner et al. [4] in 1967. The significance of Russell's discovery was then fully appreciated. It was discovered that many phenomena in physics, electronics, and biology can be described by a mathematical and physical theory of "soliton." For a historical account of the scientific development of solitons, the reader is referred to the "*Encyclopedia of Complexity and Systems Science*," especially [5, 6]. Some analytical methods leading to our present state of the art are available in several review articles [7–9].

2. Basic Properties of Solitary Solutions and Compactons [5, 6]

A soliton is a special traveling wave that after a collision with another soliton eventually emerges unscathed. Solitons are solutions of partial differential equations that model phenomena like water waves or waves along a weakly anharmonic mass-spring chain. A soliton is a bell-like solution as illustrated in Figure 1.

The soliton can be written in a standard form, which is

$$u(\xi) = p\,\mathrm{sech}^2(q\xi) = \frac{4p}{e^{2q\xi} + e^{-2q\xi} + 2}, \tag{2.1}$$

where $u(x,t) = u(\xi), \xi = x - ct$, and c is the wave velocity.

It is obvious that

$$\lim_{\xi\to\infty} u(\xi) = 0, \qquad \lim_{\xi\to-\infty} u(\xi) = 0. \tag{2.2}$$

The soliton has exponential tails, which are the basic character of solitary waves. This property allows the exponential function to describe its solution, see Section 5.9 for detailed discussion.

The soliton obeys a superposition-like principle: solitons passing through one another emerge unmodified, see Figure 2.

A compacton is a special solitary traveling wave that, unlike a soliton, does not have exponential tails. A compacton-like solution is a special wave solution which can be expressed by the squares of sinusoidal or cosinoidal functions.

3. Explanation of Compacton Solutions [6]

Compactons are special solitons with finite wavelength. It was Rosenau and Hyman [10] who first found compactons in 1993.

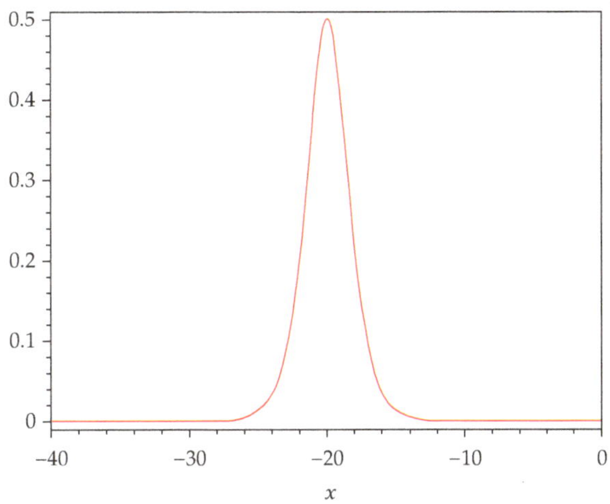

Figure 1: Bell-like solitary wave.

3.1. *Compacton: An Oscillatory Wave with No Tails*

Now consider a modified version of KdV equation in the form

$$u_t + \left(u^2\right)_x + \left(u^2\right)_{xxx} = 0. \tag{3.1}$$

Introducing a complex variable ξ defined as $\xi = x - ct$, where c is the velocity of traveling wave, integrating once, and we have

$$-cu + u^2 + \left(u^2\right)_{\xi\xi} = D, \tag{3.2}$$

where D is an integral constant, for solitary solution, and we set $D = 0$.

We rewrite (3.2) in the form

$$v_{\xi\xi} + v - cv^{1/2} = 0, \tag{3.3}$$

where $u^2 = v$.

In case $c = 0$, we have periodic solution: $v(\xi) = A\cos\xi + B\sin\xi$. Periodic solution of nonlinear oscillators can be approximated by sinusoidal function. It helps understanding if an equation can be classified as oscillatory by direct inspection of its terms.

We consider two common-order differential equations whose exact solutions are important for physical understanding:

$$u'' - k^2u = 0, \tag{3.4}$$

$$u'' + \omega^2u = 0. \tag{3.5}$$

Both equations have linear terms with constant coefficients.

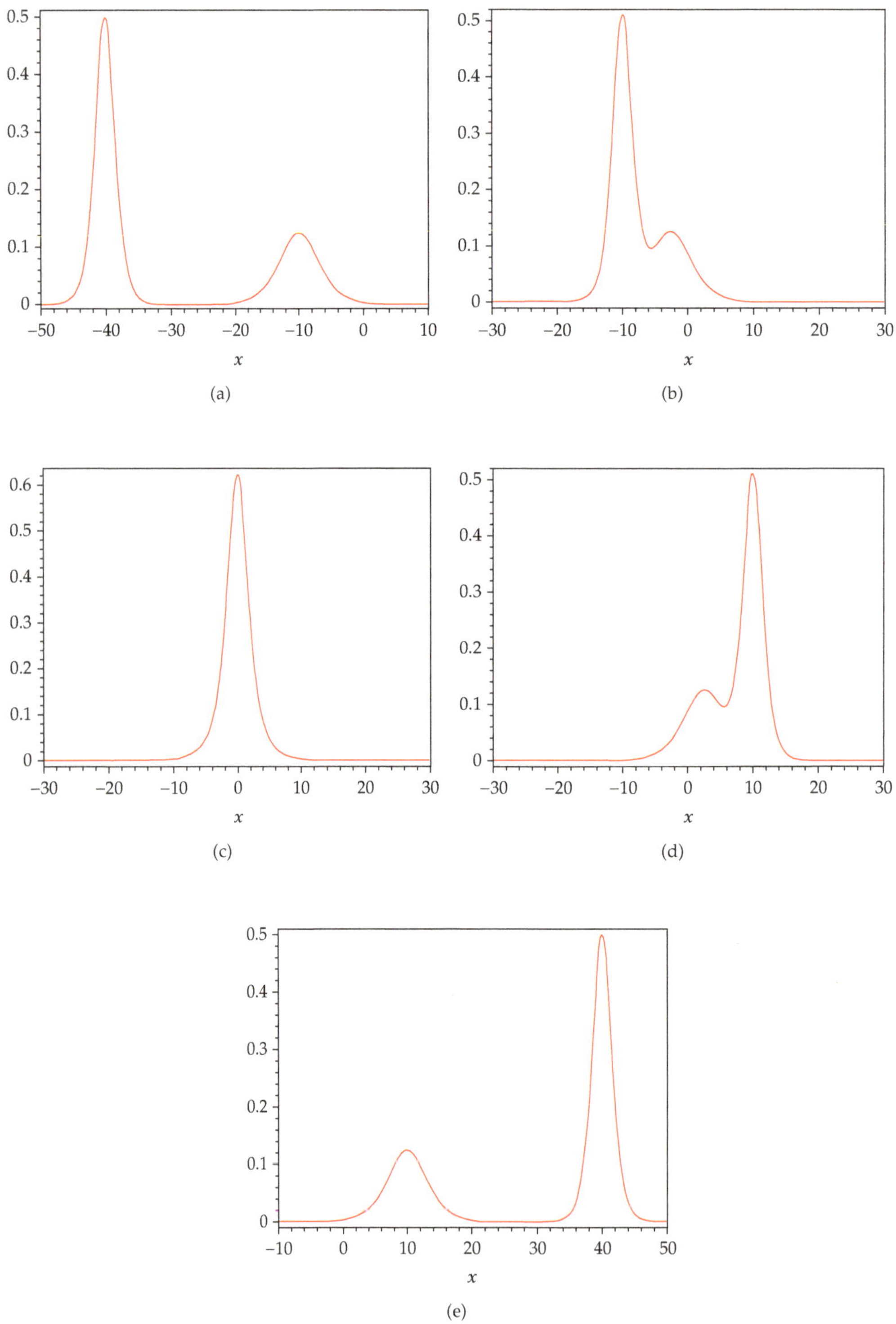

Figure 2: Collision of two solitary waves.

The crucial difference between these two very simple equations is the sign of the coefficient of u in the second term. This determines whether the solutions are exponential or oscillatory. The general solution of (3.4) is

$$u = Ae^{kt} + Be^{-kt}. \tag{3.6}$$

The second equation, (3.5), has a positive coefficient of u, and in this case, the general solution reads

$$u = A\cos\omega t + B\sin\omega t. \tag{3.7}$$

This solution describes an oscillation at the angular velocity ω.

Equation (3.3) behaves sometimes like an oscillator when $1-cv^{-1/2} > 0$, that is, $u = v^{1/2}$ has a periodic solution, and we assume that v can be expressed in the form

$$v = u^2 = A^2\cos^4\omega\xi. \tag{3.8}$$

Substituting (3.8) into (3.3) results in

$$12A^2\omega^2\cos^2\omega\xi - 16A^2\omega^2\cos^4\omega\xi - A^2\cos^4\omega\xi - cA\cos^2\omega\xi = 0. \tag{3.9}$$

We, therefore, have

$$\begin{aligned} &12A^2\omega^2 - cA = 0, \\ &-16A^2\omega^2 - A^2 = 0. \end{aligned} \tag{3.10}$$

Solving the above system, (3.10), yields

$$\omega = \frac{1}{4}, \qquad A = \frac{4}{3}c. \tag{3.11}$$

We obtain the solution in the form

$$u = v^{1/2} = \frac{4c}{3}\cos^2\left[\frac{1}{4}(x-ct)\right]. \tag{3.12}$$

By a careful inspection, v can tend to a very small value or even zero, and as a result, $1-cv^{-1/2}$ tends to negative infinite, and (3.3) behaves like (3.4) with $k \to \infty$; the exponential tails vanish completely at the edge of the bell shape (see Figure 3):

$$u = \begin{cases} \dfrac{4c}{3}\cos\left[\dfrac{1}{4}(x-ct)\right], & |x-ct| \le 2\pi, \\ 0, & \text{otherwise}. \end{cases} \tag{3.13}$$

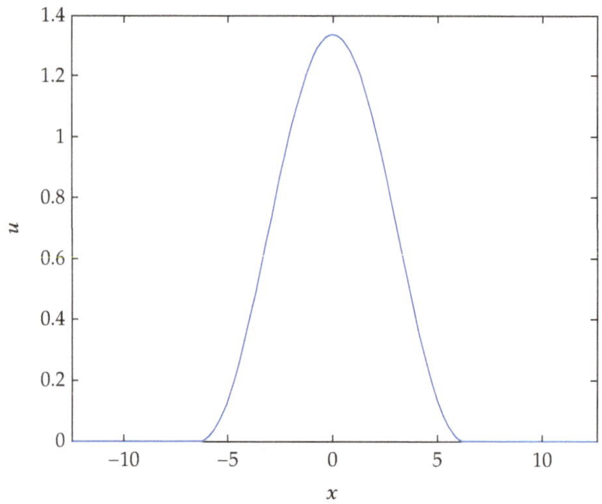

Figure 3: Compacton wave without tails.

This is a compact wave. Unlike solitons, compacton does not have exponential tails (Figure 3).

3.2. A Criterion for Oscillatory Thermopower Waves

Thermal conduction in fuel/Bi_2Te_3/Al_2O_3 or fuel/Bi_2Te_3/terracotta systems always results in strong oscillation of the output signals. A criterion for oscillatory thermopower waves is much needed.

Recently, Walia et al. proposed a theory of thermopower wave oscillations to describe coupled thermal waves in fuel/Bi_2Te_3/Al_2O_3 or fuel/Bi_2Te_3/terracotta systems [11]. The dimensionless governing equations are as follows [11]:

$$\frac{\partial u}{\partial t} = \frac{\partial^2 u}{\partial x^2} + we^{-1/u} - \ell(u - u_a), \tag{3.14}$$

$$\frac{\partial w}{\partial t} = -\beta we^{-1/u}, \tag{3.15}$$

where u is nondimensional temperature, w is the concentration of the fuel, β and ℓ are, respectively, parameters related to the properties of the fuel and volumetric heat transfer, and u_a is the ambient temperature.

Ignoring the nonlinear term in (3.14), we would have a wave solution. Changing the parameters in the system will result in strong oscillation [11], and an analytical criterion to predict oscillatory thermopower waves is very useful for design of Bi_2Te_3 films.

The system, (3.14) and (3.15), is difficult to solve analytically because of strong nonlinearity. In order to obtain a criterion for oscillatory thermopower waves, some necessary approximations are needed. Equation (3.15) is approximately written in the form

$$\frac{\partial w}{\partial t} = -\beta w\left(1 - \frac{1}{u}\right). \tag{3.16}$$

u in (3.16) is assumed to be a known function; solving w in (3.16) results in

$$w = \exp\left\{-\beta\left(1 - \frac{1}{u}\right)\right\}. \tag{3.17}$$

The nonlinear term, $we^{-1/u}$, in (3.14) is expressed in an approximate form

$$\begin{aligned} we^{-1/u} &= \exp\left\{-\beta\left(1 - \frac{1}{u}\right)\right\}e^{-1/u} = \left(1 - \beta\left(1 - \frac{1}{u}\right)\right)\left(1 - \frac{1}{u}\right) \\ &= 1 - \beta + \frac{2\beta - 1}{u} - \frac{\beta}{u^2}. \end{aligned} \tag{3.18}$$

Equation (3.14) is rewritten in the following equivalent form:

$$\frac{\partial u}{\partial t} = \frac{\partial^2 u}{\partial x^2} + 1 - \beta + \frac{2\beta - 1}{u} - \ell(u - u_a) + f(u, w), \tag{3.19}$$

where $f(u, w)$ is defined as

$$f(u, w) = we^{-1/u}\left\{1 - \beta + \frac{2\beta - 1}{u}\right\}. \tag{3.20}$$

In order to solve thermopower waves, we make a transform

$$\xi = x - ct, \tag{3.21}$$

where c is wave speed.

By the transform, (3.21), we convert (3.19) into an ordinary differential equation, which is

$$u'' + cu' + \frac{2\beta - 1}{u} - \ell u + (\ell u_a + 1 - \beta) + f(u, w) = 0. \tag{3.22}$$

We rewrite (3.22) in the form

$$u'' + cu' + \frac{2\beta - 1}{u} - \ell u = F(u, w), \tag{3.23}$$

where

$$F(u, w) = -(\ell u_a + 1 - \beta) - f(u, w). \tag{3.24}$$

Note. Equation (3.23) is exactly equivalent to (3.14). In order to solve (3.23) approximately, we write an iteration formulation

$$u''_{n+1} + cu'_{n+1} + \frac{2\beta - 1}{u_{n+1}} - \ell u_{n+1} = F(u_n(\xi), w_n(\xi)). \tag{3.25}$$

In (3.25), F can be considered as a known function of ξ. Equation (3.25) is, therefore, similar to a forced nonlinear oscillator.

We search for a periodic solution of (3.25). To this end, we assume that its solution can be expressed in the form

$$u_{n+1}(\xi) = A\cos(\omega\xi + \theta_0). \tag{3.26}$$

By an analytical method [7], we can obtain the following approximate frequency:

$$\omega = \sqrt{\frac{2(2\beta - 1)}{A^2} - \ell}. \tag{3.27}$$

The assumption, (3.26), follows $\omega > 0$, that is,

$$\frac{2(2\beta - 1)}{A^2} - \ell > 0. \tag{3.28}$$

This is a criterion for oscillatory thermopower waves. When $2(2\beta - 1)/A^2 - \ell \le 0$, we can predict thermopower waves without oscillation.

3.3. *A Criterion for Gaseous Emission Waves*

Lin and Hildemann [12] developed a general mathematical model to predict emissions of volatile organic compounds (VOCs) from hazardous or sanitary landfills. The model includes important mechanisms occurring in unsaturated subsurface landfill environments: biogas flow, leachate flow, diffusion, adsorption, degradation, volatilization, and mass transfer limitations through the top cover. Lin-Hildemann equation for gaseous emission can be expressed as follows [12]:

$$\frac{\partial u}{\partial t} + V\frac{\partial u}{\partial z} = D\frac{\partial^2 u}{\partial z^2} - \mu u, \tag{3.29}$$

with the following boundary/initial conditions:

$$Vu - D\frac{\partial u}{\partial z} = -Hu \quad \text{at } z = 0, \tag{3.30}$$

$$u(\infty, t) = 0, \tag{3.31}$$

$$u(z, 0) = A\exp(\beta z), \tag{3.32}$$

where u is the total chemical concentration per unit volume of soil, V is the effective emission speed, and D is effective diffusion coefficient. The definitions of V and D are given in [12].

This paper aims at a wave solution of (3.29). By the wave transform,

$$\xi = z - ct. \tag{3.33}$$

Equation (3.29) is converted into an ordinary differential equation, which is

$$Du'' + (c + V)u' - \mu u = 0, \tag{3.34}$$

where c is the wave speed.

Considering the boundary condition, (3.31), the solution of (3.34) is

$$u = A\, \exp(a\xi) = A\, \exp\{a(z - ct)\}, \tag{3.35}$$

where

$$a = \frac{V - c - \sqrt{(V - c)^2 + 4\mu D}}{2D}. \tag{3.36}$$

By the boundary condition, (3.30), we have

$$V - Da = -H. \tag{3.37}$$

Solving a and c from (3.36) and (3.37) results in

$$a = \frac{V + H}{D}, \tag{3.38}$$

$$c = H + \frac{\mu D}{V + H}. \tag{3.39}$$

Considering the initial condition, we have

$$\beta = \frac{V + H}{D} < 0. \tag{3.40}$$

Equation (3.40) can be written in the form

$$\beta = \frac{-v_G^B / R_G + v_L^B / R_L + K_T / R_G}{D_G^E / R_G + D_L^E / R_L} < 0, \tag{3.41}$$

where v_G^B, v_L^B are the bulk (apparent) gas and water velocities, respectively. D_G^E, D_L^E are the effective gaseous and aqueous diffusion coefficients in soil, respectively; R_G, R_L are phase-partitioning coefficients of gas and liquid, respectively.

Equation (3.41) is the criterion for gaseous emission waves.

For a wave solution, the initial condition should be expressed in the form of (3.32). If the initial condition cannot be expressed in an exponential function, the criterion for gaseous emission waves becomes invalid. The present criterion can easily be extended to various nonlinear cases.

4. Exact Solutions versus Asymptotic Solutions

There is plainly a tendency in the modern nonlinear science community to obtain exact solutions for nonlinear equations. There are many results on the exact solutions of nonlinear equations where the initial or boundary conditions are not considered. These solutions are called *mathematical solutions* because the physical constraints on the real-world problem that is being modeled are not accounted for. Our main aim, however, is to find solutions of the underlying problem that satisfy all the initial/boundary conditions that exist. These solutions, naturally, are called the physical solutions of the problem. Consider, for example, the well-known KdV equation

$$\frac{\partial u}{\partial t} - 6u\frac{\partial u}{\partial x} + \frac{\partial^3 u}{\partial x^3} = 0, \tag{4.1}$$

with a solitary solution

$$u = -\frac{c}{2}\mathrm{sech}^2\sqrt{\frac{c}{4}}(x - ct - \xi_0). \tag{4.2}$$

Many mathematical solutions for (4.1) could be found that carry no physical meaning ($u = 1$, for instance, is an exact solution of (4.1) that has no physical meaning at all). Other researchers, on the other hand, begin with some very good initial conditions, say

$$u(x,0) = -\frac{c}{2}\mathrm{sech}^2\sqrt{\frac{c}{4}}(x - \xi_0), \tag{4.3}$$

and find that the condition is in fact too good to solve the equation. For a travelling solution, for example, we might guess a solution of the form

$$u(x,t) = -\frac{c}{2}\mathrm{sech}^2\sqrt{\frac{c}{4}}(x + at - \xi_0), \tag{4.4}$$

where the unknown constant a can be determined by substituting (4.4) into (4.1).

An asymptotic approach is, however, to search for an asymptotic solution with physical understanding. If, for example, we feel interest in a solitary solution of (4.1), then we can assume that the solution has the form

$$u(x,t) = \frac{1}{a\exp\{k(x - ct)\} + b\exp\{-k(x - ct)\}}, \tag{4.5}$$

or

$$u(x,t) = \frac{1}{a\exp\{k(x-ct)\} + b\exp\{-k(x-ct)\} + d}, \tag{4.6}$$

where a, b, c, d, and k are unknown constants which can be determined via various methods.

For N-solitary solutions, we can assume that the solution has the following form:

$$u(x,t) = \sum_{i=1}^{N} \frac{1}{a_i\exp\{k_i(x-c_it)\} + b_i\exp\{-k_i(x-c_it)\} + d_i}, \tag{4.7}$$

or

$$u(x,t) = \frac{\sum_n a_n\exp(q_nx + p_nt)}{\sum_m b_m\exp(\alpha_mx + \beta_mt)}, \tag{4.8}$$

where a_i, b_i, c_i, d_i, and k_i and α_i, β_i, q_i, and p_i are unknown constants to be further determined.

For a two-wave solution, we can assume that

$$u = \frac{a_{-1}\exp(-\xi) + b_1\exp(\xi) + a_0 + b_{-1}\exp(-\eta) + b_1\exp(\eta)}{c_{-1}\exp(-\xi) + c_1\exp(\xi) + b_0 + d_{-1}\exp(-\eta) + d_1\exp(\eta)}, \tag{4.9}$$

or

$$u = \frac{1}{c_{-1}\exp(-\xi) + c_1\exp(\xi) + b_0 + d_{-1}\exp(-\eta) + d_1\exp(\eta)}, \tag{4.10}$$

where $\xi = k_1x + w_1t, \eta = k_2x + w_2t$.

Some asymptotic methods are easy and accessible to all nonmathematicians using only pencil and paper. Consider a nonlinear differential equation for corneal shape [13]

$$h'' - ah + \frac{b}{\sqrt{1+h'^2}} = 0, \tag{4.11}$$

with boundary conditions $h(1) = 0$ and $h'(0) = 0$.

Hereby we suggest a Taylor series method to find an asymptotic solution [14].

We rewrite (4.11) in the form

$$h'' = ah - b\left(1 + h'^2\right)^{-1/2}. \tag{4.12}$$

Incorporating the boundary condition, $h'(0) = 0$, we have

$$h''(0) = ah(0) - b\left(1 + h'^2(0)\right)^{-1/2} = ah_0 - b. \tag{4.13}$$

Differentiating (4.12) with respect to x results in

$$h''' = ah' + b\left(1 + h'^2\right)^{-3/2} h'h''. \tag{4.14}$$

This yields

$$h'''(0) = 0. \tag{4.15}$$

Proceeding a similar way as above, we have

$$h^{(4)}(0) = ah''(0) + bh''^2(0) = a(ah_0 - b) + b(ah_0 - b)^2. \tag{4.16}$$

Applying the Taylor series, we obtain

$$h(x) = h(0) + h'(0)x + \frac{1}{2!}h''(0)x^2 + \frac{1}{3!}h'''(0)x^3 + \frac{1}{4!}h^{(4)}(0)x^4, \tag{4.17}$$

or

$$h(x) = h_0 + \frac{1}{2}(ah_0 - b)x^2 + \frac{1}{24}\left[a(ah_0 - b) + b(ah_0 - b)^2\right]x^4. \tag{4.18}$$

Incorporating the boundary condition, $h(1) = 0$, yields

$$h(1) = h_0 + \frac{1}{2}(ah_0 - b) + \frac{1}{24}\left[a(ah_0 - b) + b(ah_0 - b)^2\right] = 0, \tag{4.19}$$

or

$$a^2bh_0^2 + \left(24 + 12a + a^2 - 2ab^2\right)h_0 - 12b - ab + b^3 = 0. \tag{4.20}$$

From (4.20), h_0 can be solved, which reads [14]

$$h_0 = \frac{-(24 + 12a + a^2 - 2ab^2) + \sqrt{(24 + 12a + a^2 - 2ab^2)^2 + 4a^2b(12b + ab - b^3)}}{2a^2b}. \tag{4.21}$$

To compare with Okrasiński and Płociniczak's result, setting $a = b = 1$, we have

$$h_0 = \frac{-35 + \sqrt{35^2 + 48}}{2} = 0.33956, \tag{4.22}$$

which is very close to Okrasiński and Płociniczak's result [13].

The accuracy can be further improved if the solution procedure continues.

Comparing the Okrasiński and Płociniczak's method with our pencil-and-paper method, we conclude that the solution process is accessible to nonmathematicians to solve any nonlinear two-point boundary problems.

5. Asymptotic Methods for Solitary Solutions

The investigation of soliton solutions of nonlinear evolution equations plays an important role in the study of nonlinear physical phenomena. There are many analytical approaches to the search for solitary solutions, see [5, 6]. Among various methods, the homotopy perturbation method, the variational iteration method, the exp-function method, and the variational approach have been worked out over a number of years by numerous authors, and they have matured into relatively fledged analytical methods for nonlinear equations thanks to the efforts of many researchers.

5.1. Soliton Perturbation

We consider the following perturbed nonlinear evolution equation [5, 15]:

$$u_T + L(u) + N(u) = \varepsilon R(u), \quad 0 < \varepsilon \ll 1. \tag{5.1}$$

When $\varepsilon = 0$, we have an unperturbed equation

$$u_T + L(u) + N(u) = 0, \tag{5.2}$$

which is assumed to have a solitary solution.

In case $\varepsilon \neq 0$, but $0 < \varepsilon \ll 1$, we can use perturbation theory and look for approximate solutions of (5.1) which are close to solitary solutions of (5.2).

Using multiple time scales (a slow time τ and fast time t scale such that $\partial_T = \partial_t + \varepsilon \partial_\tau$), we assume that the soliton solution can be expressed in the form [5, 15]

$$u(x,T) = u_0(\xi,\tau) + \varepsilon u_1(\xi,\tau,t) + \varepsilon^2 u_2(\xi,\tau,t) + \cdots, \tag{5.3}$$

where $\xi = x - ct$, τ is a slow time, and t is a fast time.

Substituting (5.3) into (5.1), then equating like powers of ε, we can obtain a series of linear equations for $u_i (i = 0, 1, 2, 3, \ldots)$, which can be solved sequentially.

In most cases, the nonlinear term $R(u)$ in (5.1) plays an import role in understanding various solitary phenomena, and the coefficient ε is not limited to "small parameter."

5.2. Modified Multitime Expansions [16]

In order to overcome the shortcoming arising in the above solution process, hereby we applied the modified multitime expansions (see Section 2.9 of [16]). To illustrate the method, we consider the following equation:

$$\frac{\partial u}{\partial T} + au\frac{\partial u}{\partial x} + b\frac{\partial^3 u}{\partial x^3} + \varepsilon R(u) = 0. \tag{5.4}$$

Introducing the time scales $T_n = \varepsilon^n t (n = 0, 1, 2, 3, \ldots)$, and using the parameter-expansion method (see Section 5.8), we assume that the solution and the constants a and b can be expressed as [5, 16]

$$u = u_0(T_0, T_1, T_2, \ldots) + \varepsilon u_1(T_0, T_1, T_2, \ldots) + \varepsilon^2 u_2(T_0, T_1, T_2, \ldots) + \cdots, \tag{5.5}$$

$$a = a_0 + \varepsilon a_1 + \varepsilon^2 a_2 + \cdots, \tag{5.6}$$

$$b = b_0 + \varepsilon b_1 + \varepsilon^2 b_2 + \cdots, \tag{5.7}$$

where the constants a_i and b_i can be identified by means of no secular terms. Hereby, we define the secular term in a more general form that the term involves time in the form $t^n f$, even in case $t^n f$ tends to zero when $t \to \infty$.

The equation for u_0 is

$$\frac{\partial u_0}{\partial t} + a_0 u_0 \frac{\partial u_0}{\partial x} + b_0 \frac{\partial^3 u_0}{\partial x^3} = 0. \tag{5.8}$$

We can choose suitably the values of a_0 and b_0, so that the solution of (5.8) can be easily obtained, and involves the basic properties of the original solution.

We use the Duffing equation to illustrate the solution procedure [16]

$$u'' + 1 \cdot u + \varepsilon u^3 = 0, \qquad u(0) = A, \qquad u'(0) = 0. \tag{5.9}$$

Suppose that the solution can be expressed in (5.5), and the coefficient, 1, can be expanded into

$$1 = \omega^2 + \varepsilon \omega_1 + \varepsilon^2 \omega_2 + \cdots. \tag{5.10}$$

Substituting (5.5) and (5.10) into (5.9) and collecting terms of the same powers of ε, we have

$$D_0^2 u_0 + \omega^2 u_0 = 0, \tag{5.11}$$

$$D_0^2 u_1 + \omega^2 u_1 + \omega_1 u_0 + 2D_0 D_1 u_0 + u_0^3 = 0, \tag{5.12}$$

where $D_i = \partial^2 / \partial T_i^2$.

Solving (5.11) with the initial conditions $u_0(0) = A$ and $u_0'(0) = 0$, we have

$$u_0 = A(T_1, T_2) \cos \omega T_0. \tag{5.13}$$

Substituting u_0 into (5.12) results in

$$D_0^2 u_1 + \omega^2 u_1 + A\left(\omega_1 + \frac{3}{4}A^2\right) \cos \omega T_0 - 2\omega D_1(A) \sin \omega T_0 + \frac{1}{4}A^3 \cos 3\omega T_0 = 0. \tag{5.14}$$

Eliminating secular terms needs

$$D_1(A) = 0,$$
$$\omega_1 = -\frac{3}{4}A^2. \tag{5.15}$$

If only the first-order approximate solution is searched for, from (5.10), we have

$$1 = \omega^2 + \varepsilon\omega_1 = \omega^2 - \frac{3}{4}\varepsilon A^2, \tag{5.16}$$

or

$$\omega = \sqrt{1 + \frac{3}{4}\varepsilon A^2}. \tag{5.17}$$

The obtained frequency-amplitude relationship, (5.17), is valid for the whole solution domain, and the maximal relative error is less than 7% when $\varepsilon A^2 \to \infty$.

5.3. Variational Approach

This section is an elementary introduction to the concepts of the calculus of variations and its applications to solitary solutions. Generally speaking, there exist two basic ways to describe a nonlinear problem: (1) by differential equations (DE) with initial/boundary conditions; (2) by variational principles (VP). The former is widely used, while the later is rarely used in solitary theory. The VP model has many advantages over its DE partner: simple and compact in form while comprehensive in content, encompassing implicitly almost all information characterizing the problem under consideration. Variational methods have been, and continue to be, popular tools for nonlinear problems. When contrasted with other approximate analytical methods, variational methods combine the following two advantages: (1) they provide physical insight into the nature of the solution of the problem; (2) the obtained solutions are the best among all the possible trial functions.

5.3.1. Inverse Problem of Calculus of Variations

The inverse problem of calculus of variations is to establish a variational formulation directly from governing equations and boundary/initial conditions. We will use the semi-inverse method [17, 18] to establish various variational principles directly from the governing equations.

Consider the well-known Korteweg-de Vries (KdV) equation

$$u_t + auu_x + bu_{xxx} = 0, \tag{5.18}$$

where a and b are constants, and the subscripts denote partial differentiations.

We rewrite it in a conserved form

$$\frac{\partial u}{\partial t}+\frac{\partial}{\partial x}\left(\frac{1}{2}au^2+bu_{xx}\right)=0. \tag{5.19}$$

According to the conservation form of (5.19), we can introduce a potential functional Φ defined by

$$\Phi_x = u, \tag{5.20}$$

$$\Phi_t = -\left(\frac{1}{2}au^2+bu_{xx}\right). \tag{5.21}$$

So the KdV equation can be written in the form

$$\Phi_{xt}+a\Phi_x\Phi_{xx}+b\Phi_{xxxx}=0, \tag{5.22}$$

which can be derived from the following variational principle using the semi-inverse method [17]:

$$J(\Phi)=\iint\left\{\frac{1}{2}\Phi_x\Phi_t+\frac{a}{6}\Phi_x^3-\frac{b}{2}\Phi_{xx}^2\right\}dt\,dx. \tag{5.23}$$

In order to obtain a generalized variational principle with two independent fields (Φ, u), we apply the Lagrange multiplier to (5.23)

$$J(\Phi,u,\lambda)=\iint\left\{\frac{1}{2}\Phi_x\Phi_t+\frac{a}{6}\Phi_x^3-\frac{b}{2}\Phi_{xx}^2+\lambda(\Phi_x-u)\right\}dt\,dx. \tag{5.24}$$

The stationary condition with respect to u results in

$$\lambda = 0. \tag{5.25}$$

The Lagrange multiplier method is not valid for the case. This phenomenon is called Lagrange crisis. Hereby, we suggest three ways to overcome the crisis [18].

(1) *The Semi-Inverse Method [17, 18]*

Generally, the multiplier can be expressed in the form after identification

$$\lambda=\lambda(u,\Phi,u_t,u_x,\Phi_t,\Phi_x,\ldots). \tag{5.26}$$

We replace the last term including the Lagrange multiplier by a new variable F, that is,

$$J(\Phi,u)=\iint\left\{\frac{1}{2}\Phi_x\Phi_t+\frac{a}{6}\Phi_x^3-\frac{b}{2}\Phi_{xx}^2+F\right\}dt\,dx, \tag{5.27}$$

where F is unknown, which can be expressed in the form

$$F = F(u, \Phi, u_t, u_x, \Phi_t, \Phi_x, \ldots). \tag{5.28}$$

Equation (5.27) is such constructed according to the semi-inverse method [17]. To identify F, making the functional stationary with respect to u, we have

$$\frac{\delta F}{\delta u} = 0, \tag{5.29}$$

where $\delta F / \delta u$ is the variational derivative defined as

$$\frac{\delta F}{\delta u} = \frac{\partial F}{\partial u} - \frac{\partial}{\partial t}\left(\frac{\partial F}{\partial u_t}\right) - \frac{\partial}{\partial x}\left(\frac{\partial F}{\partial u_x}\right) + \frac{\partial^2}{\partial x^2}\left(\frac{\partial F}{\partial u_{xx}}\right) + \cdots . \tag{5.30}$$

Equation (5.29) should be equivalent to (5.20), and this requires

$$\frac{\delta F}{\delta u} = c(\Phi_x - u), \tag{5.31}$$

where c is a nonzero constant. From (5.31), we can determine F as follows:

$$F = -\frac{1}{2}c(\Phi_x - u)^2 = \tilde{c}(\Phi_x - u)^2, \tag{5.32}$$

where $\tilde{c}$ is a constant, $\tilde{c} = -c/2$.

We, therefore, obtain the following needed variational principle:

$$J(\Phi, u) = \iint \left\{ \frac{1}{2}\Phi_x\Phi_t + \frac{a}{6}\Phi_x^3 - \frac{b}{2}\Phi_{xx}^2 + \tilde{c}(\Phi_x - u)^2 \right\} dt\, dx. \tag{5.33}$$

Its Euler-Lagrange equations are

$$\begin{gathered} -2\tilde{c}(\Phi_x - u) = 0, \\ -\Phi_{xt} - a\Phi_x\Phi_{xx} - b\Phi_{xxxx} - 2\tilde{c}(\Phi_x - u)_x = 0, \end{gathered} \tag{5.34}$$

which satisfy the field equations (5.20) and (5.22), respectively.

(2) *The Hidden Lagrange Multiplier [18]*

Let us come back to (5.25), $\lambda = 0$, which should be the constraint equation. This means that (5.25) inexplicitly involves a lost constraint equation, so we can identify the multiplier in the form [18]

$$\lambda = \tilde{c}(\Phi_x - u). \tag{5.35}$$

This results in the same result above.

(3) *Replacement of Some Variables in the Original Functional Using the Constraint Equation*

Sometimes the Lagrange crisis can be eliminated by replacing some variables in the original variational principle using the constraint equation; a detailed discussion was systematically given in [18].

We replace $\Phi_x\Phi_t$ by $u\Phi_t$ in (5.23) and introduce a Lagrange multiplier in the resultant function

$$J(\Phi, u, \lambda) = \iint \left\{ \frac{1}{2}u\Phi_t + \frac{a}{6}\Phi_x^3 - \frac{b}{2}\Phi_{xx}^2 + \lambda(\Phi_x - u) \right\} dt\, dx. \tag{5.36}$$

Identification of the multiplier yields

$$\lambda = \frac{1}{2}\Phi_t. \tag{5.37}$$

Submitting the identified multiplier into (5.36) results in

$$\begin{aligned} J(\Phi, u) &= \iint \left\{ \frac{1}{2}u\Phi_t + \frac{a}{6}\Phi_x^3 - \frac{b}{2}\Phi_{xx}^2 + \frac{1}{2}\Phi_t(\Phi_x - u) \right\} dt\, dx \\ &= \iint \left\{ \frac{a}{6}\Phi_x^3 - \frac{b}{2}\Phi_{xx}^2 + \frac{1}{2}\Phi_t\Phi_x \right\} dt\, dx. \end{aligned} \tag{5.38}$$

We find that the constraint is not eliminated yet, and this is another Lagrange crisis, which can also be eliminated by the semi-inverse method

$$J(\Phi, u) = \iint \left\{ \frac{1}{2}u\Phi_t + \frac{a}{6}\Phi_x^3 - \frac{b}{2}\Phi_{xx}^2 + F \right\} dt\, dx. \tag{5.39}$$

The stationary conditions with respect to u and Φ are, respectively, as follows:

$$\begin{aligned} \frac{1}{2}\Phi_t + \frac{\delta F}{\delta u} &= 0, \\ -\frac{1}{2}u_t - a\Phi_x\Phi_{xx} - b\Phi_{xxxx} + \frac{\delta F}{\delta \Phi} &= 0. \end{aligned} \tag{5.40}$$

According to (5.20) and (5.21), we have

$$\begin{aligned} \frac{\delta F}{\delta u} &= -\frac{1}{2}\Phi_t = \frac{1}{4}au^2 + \frac{1}{2}bu_{xx}, \\ \frac{\delta F}{\delta \Phi} &= \frac{1}{2}u_t + u\Phi_x\Phi_{xx} + b\Phi_{xxxx} = \frac{1}{2}u_t - \Phi_{xt} = \frac{1}{2}\Phi_{xt} - \Phi_{xt} = -\frac{1}{2}\Phi_{xt}. \end{aligned} \tag{5.41}$$

From (5.41), F can be determined as

$$F = \frac{1}{12}au^3 - \frac{1}{4}b(u_x)^2 + \frac{1}{4}\Phi_x\Phi_t. \tag{5.42}$$

We, therefore, obtain the following variational principle:

$$J(\Phi, u) = \iint \left\{ \frac{1}{2}u\Phi_t + \frac{a}{6}\Phi_x^3 - \frac{b}{2}\Phi_{xx}^2 + \frac{1}{12}au^3 - \frac{1}{4}b(u_x)^2 + \frac{1}{4}\Phi_x\Phi_t \right\} dt\,dx. \tag{5.43}$$

By the semi-inverse method, we can obtain various different two-field variational principles, and we write here the following one for reference:

$$J(\Phi, u) = \iint \left\{ \frac{1}{2}\Phi_x\Phi_t + a\left(\frac{1}{3}u^3 - \frac{1}{2}u^2\Phi_x\right) - \frac{b}{2}\Phi_{xx}^2 \right\} dt\,dx. \tag{5.44}$$

The potential in the above functionals (5.23), (5.33), and (5.44) requires second order of differentiation (Φ_{xx}), leading to the complications in the finite element calculation. For the purpose of simplification in finite element computation, we often introduce some additional variables to reduce the order of differentiations. This is of course equivalent to introducing some additional constraints in the variational principle. Generally, we can eliminate the introduced constraints by the Lagrange multiplier method, but as illustrated above, the method might fail.

Now we introduce a new variable E defined as

$$E = \frac{a}{2}u^2 + bu_{xx}. \tag{5.45}$$

By the semi-inverse method [17, 18], we obtain the following three-field variational principle:

$$J(\Phi, u, E) = \iint \left\{ u\Phi_t + E\Phi_x + \frac{a}{3}u^3 - bu_x^2 - Eu \right\} dt\,dx. \tag{5.46}$$

It is obvious that all variables in the obtained functional (5.46) are in first-order differentiations, leading to much convenience in numerical simulation.

It is easy to establish a variational formulation by introducing a potential function, and we can also establish a variational principle without auxiliary special function. To elucidate this, we consider the KdV equation in the form

$$u_t + uu_x + u_{xxx} = 0, \tag{5.47}$$

where subscripts denote partial differentiations. If we introduce a velocity potential Φ defined as $u = \Phi_x$, then the KdV equation can be derived from the variational principle

$$\delta \iint L\,dt\,dx = 0, \tag{5.48}$$

where the Lagrangian L can be expressed in the form

$$L = \frac{1}{2}\Phi_x\Phi_t + \frac{1}{6}\Phi_x^3 - \frac{1}{2}\Phi_{xx}^2. \tag{5.49}$$

Our aim is to search for a Lagrangian for (5.47). It is easy to establish a variational formulation for differential equations with even orders. The KdV equation has odd-order differentiations, and therefore, no Lagrangian for (5.47). To circumvent this obstacle, we take partial differentiation with respect to x to both sides of KdV equation, which turns out to be the following form:

$$u_{tx} + u_x^2 + uu_{xx} + u_{xxxx} = 0. \tag{5.50}$$

By the semi-inverse method, we construct a trial Lagrangian in the form

$$L = -\frac{1}{2}u_xu_t + \frac{1}{2}u_{xx}^2 + puu_x^2 + qu^2u_{xx}, \tag{5.51}$$

where p and q are constants to be further determined. Its Euler equation can be readily obtained as follows:

$$u_{tx} + u_{xxxx} + p\left[u_x^2 - 2(uu_x)_x\right] + q\left[2uu_{xx} - \left(u^2\right)_{xx}\right] = 0, \tag{5.52}$$

or

$$u_{tx} + u_{xxxx} + (-p - 2q)u_x^2 - 2puu_{xx} = 0. \tag{5.53}$$

Setting $p = -1/2$, $q = -1/4$, then (5.53) turns out to be the modification version of the KdV equation (5.50).

Finally, we have the following needed Lagrangian in the form of velocity:

$$L = -\frac{1}{2}u_xu_t + \frac{1}{2}u_{xx}^2 - \frac{1}{2}uu_x^2 - \frac{1}{4}u^2u_{xx}. \tag{5.54}$$

This approach can be extended to many nonlinear equations. Consider the modified KdV equation

$$u_t + mu^2u_x + nu_{xxx} = 0. \tag{5.55}$$

Similarly, we change the equation so that it has even-order differentiations

$$u_{tx} + 2muu_x^2 + mu^2u_{xx} + nu_{xxxx} = 0. \tag{5.56}$$

By the same manipulation as illustrated above, we construct a trial Lagrange function in the form

$$L=-\frac{1}{2}u_x u_t+\frac{1}{2}nu_{xx}^2+pu^2u_x^2+qu^3u_{xx}, \tag{5.57}$$

where p and q are constants to be further determined. Its Euler equation can be readily obtained as follows:

$$u_{tx}+nu_{xxxx}+p\left[2uu_x^2-2\left(u^2u_x\right)_x\right]+q\left[3u^2u_{xx}-\left(u^3\right)_{xx}\right]=0, \tag{5.58}$$

or

$$u_{tx}+nu_{xxxx}+p\left[2uu_x^2-2\left(2uu_x^2+u^2u_{xx}\right)\right]+q\left[3u^2u_{xx}-\left(6uu_x^2+3u^2u_{xx}\right)\right]=0, \tag{5.59}$$

or

$$u_{tx}+nu_{xxxx}+(-2p-6q)uu_x^2-2pu^2u_{xx}=0. \tag{5.60}$$

Setting

$$p=-\frac{m}{2}, \qquad q=-\frac{m}{6}, \tag{5.61}$$

(5.60) becomes (5.56), and we, therefore, obtain the following Lagrangian:

$$L=-\frac{1}{2}u_x u_t+\frac{1}{2}nu_{xx}^2-\frac{m}{2}u^2u_x^2-\frac{m}{6}u^3u_{xx}. \tag{5.62}$$

We can also use the semi-inverse method to establish a family of variational principles for a nonlinear system. We use one-dimensional traffic flow as an example.

The research on traffic flow began at the beginning of the 20th century. Lighthill and Whitham first proposed the fluid-dynamical model for traffic flow [19, 20]. The continuum equation for unsteady one-dimensional traffic flow can be, therefore, written as

$$\frac{\partial}{\partial t}(\rho A)+\frac{\partial}{\partial x}(\rho uA)=q, \tag{5.63}$$

where A is the cross-sectional area of the road, u is the velocity, ρ is the density of cars, and q is the source. The deficiency of the model is that the traffic flow actually cannot be considered as a continuum, and to eliminate this deficiency, a fractional differential model can be introduced:

$$\frac{\partial}{\partial t}(\rho A)+\frac{D^\alpha}{Dx^\alpha}(\rho uA)=q, \quad 0<\alpha<1, \tag{5.64}$$

where D^α/Dx^α is the fractional differential, see Section 7.

In 1994, Zheng [21] suggested the following traffic flow model:

$$\frac{\partial}{\partial t}(\rho) + \frac{\partial}{\partial x}(\rho u) = 0, \tag{5.65}$$

$$\frac{\partial}{\partial t}(\rho u) + \frac{\partial}{\partial x}\left[\rho u^2 + \frac{1}{4n}(n-1)^2 u_f^2 \rho_0^{1-n} \rho^n\right] = 0, \tag{5.66}$$

where u_f is the possible maximal velocity, m is a constant, and ρ_0 is the minimal traffic density when the cars can travel at a maximal velocity.

In order to establish a variational principle for the system, we rewrite (5.66) in the following equivalent form:

$$\frac{\partial}{\partial t}(u) + \frac{\partial}{\partial x}\left[\frac{1}{2}u^2 + \frac{1}{4}(n-1)u_f^2 \rho_0^{1-n} \rho^{n-1}\right] = 0. \tag{5.67}$$

According to (5.65) and (5.67), we can introduce two functions Φ and Ψ defined as

$$\frac{\partial \Phi}{\partial x} = -\rho, \qquad \frac{\partial \Phi}{\partial t} = \rho u, \tag{5.68}$$

$$\frac{\partial \Psi}{\partial x} = -u, \qquad \frac{\partial \Psi}{\partial t} = \frac{1}{2}u^2 + \frac{1}{4}(n-1)u_f^2 \rho_0^{1-n} \rho^{n-1}, \tag{5.69}$$

so that (5.65) and (5.66) are automatically satisfied.

The essence of the semi-inverse method [17, 18] is to construct an energy-like functional with a certain unknown function, which can be identified step by step. An energy-like trial functional for the discussed problem can be constructed in the following form:

$$J(u,\rho,\Phi) = \int_0^t \int_0^L \left\{ u\frac{\partial \Phi}{\partial t} + \left[\frac{1}{2}u^2 + \frac{1}{4}(n-1)u_f^2 \rho_0^{1-n} \rho^{n-1}\right]\frac{\partial \Phi}{\partial x} + F \right\} dt\, dx, \tag{5.70}$$

where u, ρ, and Φ are considered as independent variables, and F is an unknown function of u, ρ and/or their derivatives.

There exist various approaches to the establishment of energy-like trial functionals for a physical problem, and illustrative examples can be found in [22, 23].

The advantage of the above trial functional lies on the fact that the stationary condition with respective to Φ,

$$\delta_\Phi J(u,h,\Phi) = \int_0^t \int_0^L \left\{ -\frac{\partial}{\partial t}(u) - \frac{\partial}{\partial x}\left[\frac{1}{2}u^2 + \frac{1}{4}(n-1)u_f^2 \rho_0^{1-n} \rho^{n-1}\right] \right\} \delta\Phi dt\, dx = 0, \tag{5.71}$$

leads to (5.67).

Calculating the functional (5.71) stationary with respect to u and ρ, we obtain the following Euler equations:

$$\frac{\partial \Phi}{\partial t} + u\frac{\partial \Phi}{\partial x} + \frac{\delta F}{\delta u} = 0,$$
$$\left[\frac{1}{4}(n-1)^2 u_f^2 \rho_0^{1-n} \rho^{n-2}\right]\frac{\partial \Phi}{\partial x} + \frac{\delta F}{\delta \rho} = 0. \quad (5.72)$$

We search for such an F, so that the above (5.72) satisfies the two-field equations. To this end, we set

$$\frac{\delta F}{\delta u} = -\frac{\partial \Phi}{\partial t} - u\frac{\partial \Phi}{\partial x} = -\rho u + u\rho = 0,$$
$$\frac{\delta F}{\delta \rho} = -\left[\frac{1}{4}(n-1)^2 u_f^2 \rho_0^{1-n} \rho^{n-2}\right]\frac{\partial \Phi}{\partial x} = \frac{1}{4}(n-1)^2 u_f^2 \rho_0^{1-n} \rho^{n-1}. \quad (5.73)$$

From (5.73), we can immediately identify the unknown F, which reads

$$F = \frac{1}{4n}(n-1)^2 u_f^2 \rho_0^{1-n} \rho^n. \quad (5.74)$$

So we obtain the following required variational functional:

$$J(u,\rho,\Phi) = \int_0^t \int_0^L \left\{\frac{1}{4n}(n-1)^2 u_f^2 \rho_0^{1-n} \rho^n + u\frac{\partial \Phi}{\partial t} + \left[\frac{1}{2}u^2 + \frac{1}{4}(n-1)u_f^2 \rho_0^{1-n} \rho^{n-1}\right]\frac{\partial \Phi}{\partial x}\right\} dt\, dx. \quad (5.75)$$

Proof. The Euler equations of the above functional (5.75) are

$$\delta\rho : \frac{1}{4}(n-1)^2 u_f^2 \rho_0^{1-n} \rho^{n-1} + \frac{1}{4}(n-1)^2 u_f^2 \rho_0^{1-n} \rho^{n-2}\frac{\partial \Phi}{\partial x} = 0, \quad (5.76)$$

$$\delta u : \frac{\partial \Phi}{\partial t} + u\frac{\partial \Phi}{\partial x} = 0, \quad (5.77)$$

$$\delta\Phi : -\frac{\partial}{\partial t}(u) - \frac{\partial}{\partial x}\left[\frac{1}{2}u^2 + \frac{1}{4}(n-1)u_f^2 \rho_0^{1-n} \rho^{n-1}\right] = 0. \quad (5.78)$$

From (5.68), we have $\partial\Phi/\partial x = -\rho$. Substituting the result into (5.77) leads to $\partial\Phi/\partial t = \rho u$.

From the above three-field variational functional, we can easily obtain two-field or one-field variational function by substituting one- or two-field equations into the functional (5.75). For example, substituting $\partial\Phi/\partial x = -\rho$ into (5.75), we obtain a two-field variational functional

$$J_1(u,\Phi) = \int_0^t \int_0^L \left\{\frac{1}{4n}(n-1)^2 u_f^2 \rho_0^{1-n} \rho^n + u\frac{\partial \Phi}{\partial t} - \left[\frac{1}{2}\rho u^2 + \frac{1}{4}(n-1)u_f^2 \rho_0^{1-n} \rho^n\right]\right\} dt\, dx, \quad (5.79)$$

where the variable ρ is not now an independent field. Further constraining the two-field functional (5.79) by the equation $\partial\Phi/\partial t = \rho u$, we have

$$J_2(\Phi) = \frac{1}{4n}(n-1)^2 u_f^2 \rho_0^{1-n} \int_0^t \int_0^L \rho^n dt\, dx = \int_0^t \int_0^L P dt\, dx, \tag{5.80}$$

where P is the traffic pressure defined as

$$P = C\rho^n = \frac{1}{4n}(n-1)^2 u_f^2 \rho_0^{1-n} \rho^n. \tag{5.81}$$

The functional (5.81) has the same form of the well-known Bateman principle in fluid mechanics [18].

By a paralleling operation, we can also establish a variational functional with free fields u, ρ, and Ψ. A trial functional with an unknown function F can be constructed as follows:

$$\tilde{J}(u, \rho, \Psi) = \int_0^t \int_0^L \left\{ \rho \frac{\partial \Psi}{\partial t} + \rho u \frac{\partial \Psi}{\partial x} + F \right\} dt\, dx. \tag{5.82}$$

Here, the unknown F is free from Ψ and its derivatives. By the same manipulation as illustrated above, we set

$$\begin{aligned} \frac{\delta F}{\delta u} &= -\rho \frac{\partial \Psi}{\partial x} = \rho u, \\ \frac{\delta F}{\delta \rho} &= -\frac{\partial \Psi}{\partial t} - u \frac{\partial \Psi}{\partial x} = \frac{1}{2} u^2 - \frac{1}{4}(n-1) u_f^2 \rho_0^{1-n} \rho^{n-1}. \end{aligned} \tag{5.83}$$

From (5.83), we can determine the unknown F as follows:

$$F = \frac{1}{2}\rho u^2 - \frac{1}{4n}(n-1) u_f^2 \rho_0^{1-n} \rho^n. \tag{5.84}$$

So we obtain the following needed variational principle:

$$\tilde{J}(u, \rho, \Psi) = \int_0^t \int_0^L \left\{ \frac{1}{2}\rho u^2 - \frac{1}{4n}(n-1) u_f^2 \rho_0^{1-n} \rho^n + \rho \frac{\partial \Psi}{\partial t} + \rho u \frac{\partial \Psi}{\partial x} \right\} dt\, dx. \tag{5.85}$$

It is easy to prove that the Euler equations of the above functional (5.85) satisfy the field equations (5.67) and (5.69).

Constraining the functional (5.85) by the equation $\partial\Psi/\partial x = -u$, we obtain

$$\tilde{J}_1(\rho, \Psi) = \int_0^t \int_0^L \left\{ -\frac{1}{2}\rho \left(\frac{\partial \Psi}{\partial x} \right)^2 - \frac{1}{4n}(n-1) u_f^2 \rho_0^{1-n} \rho^n + \rho \frac{\partial \Psi}{\partial t} \right\} dt\, dx, \tag{5.86}$$

which is under the constraint of the equation $\partial\Psi/\partial x = -u$. Further substituting

$$\frac{\partial\Psi}{\partial t} = \left(\frac{1}{2}\right)u^2 + \left(\frac{1}{4}\right)(n-1)u_f^2\rho_0^{1-n}\rho^{n-1} \tag{5.87}$$

into (5.86) results in

$$\widetilde{J}_2 = \int_0^t\int_0^L\left\{\frac{1}{2}\rho u^2 - \frac{1}{4n}(n-1)u_f^2\rho_0^{1-n}\rho^n\right\}dt\,dx = \int_0^t\int_0^L\left\{\frac{1}{2}\rho u^2 - P\right\}dt\,dx, \tag{5.88}$$

which is similar to the well-known Hamilton principle. □

5.3.2. A Possible Connection between the Uncertain Principle and the Least Action Principle

Maupertuis-Lagrange's principle of least kinetic potential action for a particle with mass m can be expressed as follows [18]:

$$\int_{t_1}^{t_2}\frac{1}{2}mv^2dt \longrightarrow \min. \tag{5.89}$$

We rewrite (5.89) in the form

$$\int_{s_1}^{s_2}\frac{1}{2}mv\cdot ds \longrightarrow \min, \tag{5.90}$$

or

$$\int_{s_1}^{s_2}p\cdot ds \longrightarrow \min. \tag{5.91}$$

Equation (5.91) can be approximately written in the form

$$(p_2 - p_1)\cdot ds_{12} = E_{\min}, \tag{5.92}$$

where $ds_{12} = s_2 - s_1$. Equation (5.92) means that given that the particle begins at position s_1 at time t_1 and ends at position s_2 at time t_2, the physical trajectory that connects these two endpoints is an extremum of $\Delta p\cdot\Delta s$, where Δs is the standard deviation of the displacement, and Δp is the deviation of the momentum.

For arbitrary Δs or Δp, the following inequality holds:

$$\Delta p\cdot\Delta s \geq E_{\min}. \tag{5.93}$$

This is similar to the uncertainty principle.

In optics, Fermat's principle or the principle of least time is the idea that the path taken between two points by a ray of light is the path that can be traversed in the least time. This principle is sometimes taken as the definition of a ray of light

$$t = \int_{s_1}^{s_2} \frac{ds}{\nu} \longrightarrow \min . \tag{5.94}$$

Equation (5.94) can be approximately written in the form

$$\left(\frac{1}{\nu_2} - \frac{1}{\nu_1}\right) ds_{12} \longrightarrow \min, \tag{5.95}$$

where $ds_{12} = s_2 - s_1$.

The light trajectory that connects these two endpoints s_1 and s_2 satisfies the following equation:

$$\left(\frac{1}{\nu_2} - \frac{1}{\nu_1}\right) ds = T_{\min}, \tag{5.96}$$

or

$$-\frac{1}{\nu_1 \nu_2} \Delta v \cdot \Delta s = T_{\min}, \tag{5.97}$$

where $\Delta s = s_2 - s_1$, $\Delta \nu = \nu_2 - \nu_1$.

For arbitrary Δs or Δv, the following inequality holds:

$$\begin{gathered} -\Delta v \cdot \Delta s = \nu_1 \nu_2 T_{\min} \geq c^2 T_{\min}, \\ \Delta v \cdot \Delta s \leq c^2 T_{\min}. \end{gathered} \tag{5.98}$$

5.3.3. *Variational Approach to Nonlinear Oscillators [24]*

Consider a general nonlinear oscillator in the form

$$u'' + f(u) = 0. \tag{5.99}$$

Its variational principle can be easily established as follows:

$$J(u) = \int_0^{T/4} \left\{ -\frac{1}{2} u'^2 + F(u) \right\} dt, \tag{5.100}$$

where T is the period of the nonlinear oscillator, $\partial F / \partial u = f$.

Assume that its solution can be expressed as

$$u(t) = p \cos qt, \tag{5.101}$$

where p and q are the amplitude and frequency of the oscillator, respectively.

Substituting (5.101) into (5.100) results in

$$J(p,q) = \int_0^{T/4} \left\{ -\frac{1}{2}p^2q^2\sin^2 qt + F(A\cos qt) \right\} dt. \tag{5.102}$$

Instead of setting $\partial J/\partial q = 0$ and $\partial J/\partial p = 0$, we only set [8, 24]

$$\frac{\partial J}{\partial p} = 0, \tag{5.103}$$

from which the relationship between the amplitude and frequency of the oscillator can be obtained.

Explanation of (5.103) was given in [8].

Consider a nonlinear oscillator with fractional potential [24]:

$$u'' + \varepsilon u^{1/3} = 0. \tag{5.104}$$

Its variational formulation can be readily obtained as follows:

$$J(u) = \int_0^{T/4} \left\{ -\frac{1}{2}u'^2 + \frac{3}{4}\varepsilon u^{4/3} \right\} dt. \tag{5.105}$$

Substituting (5.101) into (5.105), we obtain

$$J = \int_0^{T/4} \left\{ -\frac{1}{2}p^2q^2\sin^2 qt + \frac{3}{4}p^{4/3}\varepsilon\cos^{4/3} qt \right\} dt. \tag{5.106}$$

Setting

$$\frac{\partial J}{\partial A} = \int_0^{T/4} \left\{ -A\omega^2\sin^2\omega t + A^{1/3}\varepsilon\cos^{4/3}\omega t \right\} dt = 0, \tag{5.107}$$

we have

$$\omega^2 = \frac{\int_0^{T/4} \varepsilon\cos^{4/3}\omega t\, dt}{A^{2/3}\int_0^{T/4} \sin^2\omega t\, dt} = \frac{\int_0^{\pi/2} \varepsilon\cos^{4/3} t\, dt}{A^{2/3}\int_0^{\pi/2} \sin^2 t\, dt} = \frac{1.15959526696393\varepsilon}{A^{3/2}}. \tag{5.108}$$

The exact frequency is $\omega = 1.070451\varepsilon^{1/2}A^{-1/3}$. The 0.597% accuracy is remarkably good.

5.3.4. *Variational Approach to Chemical Reactions*

As an illustration, consider the following chemical reaction [25]:

$$nA \longrightarrow C + D, \tag{5.109}$$

which obeys the equation

$$\frac{dx}{dt} = k(a-x)^n, \qquad x(0) = 0, \tag{5.110}$$

where a is the number of molecules A at $t = 0$, x is the number of molecules C (or D) after time t, and k is a reaction constant. At the start of reaction ($t = 0$), there are no molecules C (or D) yet formed, so that the initial condition is $x(0) = 0$.

In order to obtain a variational model, we differentiate both sides of (5.110) with respect to time, resulting in

$$\frac{d^2x}{dt^2} = -kn(a-x)^{n-1}\frac{dx}{dt}. \tag{5.111}$$

Substituting (5.111) into (5.110), we obtain the following second-order differential equation:

$$\frac{d^2x}{dt^2} = -k^2n(a-x)^{2n-1}, \qquad x(0) = 0, \qquad x'(0) = ka^n, \tag{5.112}$$

which admits a variational expression in the form

$$J(x) = \int_0^\infty \left\{ \frac{1}{2}\left(\frac{dx}{dt}\right)^2 + \frac{1}{2}k^2(a-x)^{2n} \right\} dt. \tag{5.113}$$

Its Hamiltonian, therefore, can be written in the form

$$H = \frac{1}{2}\left(\frac{dx}{dt}\right)^2 - \frac{1}{2}k^2(a-x)^{2n}, \tag{5.114}$$

where H is a Hamiltonian constant, and it can be determined from initial conditions:

$$H = \frac{1}{2}\dot{x}(0)^2 - \frac{1}{2}k^2(a-x(0))^{2n} = 0. \tag{5.115}$$

Equation (5.114) becomes

$$\left(\frac{dx}{dt}\right)^2 = k^2(a-x)^{2n}. \tag{5.116}$$

In view of the initial conditions $x(0) = 0$ and $x'(0) = ka^n$, (5.116) is equivalent to (5.110). This means that the variational principle, (5.113), is exactly equivalent to its differential partner, (5.110).

Assume that the solution can be expressed in the form

$$x = a(1-e^{-\eta t}), \tag{5.117}$$

where η is an unknown constant to be further determined.

Substituting (5.117) into (5.113), and setting $dJ/d\eta = 0$, we obtain

$$\eta = \frac{1}{\sqrt{n}} k a^{n-1}. \tag{5.118}$$

So we obtain a first-order approximate solution for the discussed problem

$$x = a\left[1 - \exp\left(-n^{-1/2} k a^{n-1} t\right)\right]. \tag{5.119}$$

In order to improve accuracy, we can assume that the solution can be expressed in a more general form

$$x = a\left(1 - \sum_{i=1}^{m} b_i e^{-\eta_i t}\right), \tag{5.120}$$

which should satisfy initial conditions $x(0) = 0$, and this requires

$$1 - \sum_{i=1}^{m} b_i = 0. \tag{5.121}$$

Substituting (5.120) into (5.113), we set

$$\begin{aligned} \frac{\partial J}{\partial \eta_i} &= 0 \quad (i = 1 \sim m), \\ \frac{\partial J}{\partial b_i} &= 0 \quad (i = 2 \sim m). \end{aligned} \tag{5.122}$$

Solving (5.121)-(5.122) simultaneously, we can easily determine $2m$ parameters. The solution procedure is similar to that illustrated in [25], and we will not discuss in details to solve space.

We can also choose the following trial function:

$$x = a\left\{1 - \frac{1}{(1+pt)^q}\right\}, \tag{5.123}$$

where p and q are unknown constants to be further determined. It is obvious that (5.123) satisfies the conditions $x(0) = 0$ and $x(+\infty) = a$. Submitting (5.123) into (5.113), and setting

$$\begin{aligned} \frac{\partial J}{\partial p} &= 0, \\ \frac{\partial J}{\partial q} &= 0, \end{aligned} \tag{5.124}$$

we obtain with ease

$$p = (n-1)ka^{n-1}, \qquad q = \frac{1}{n-1}. \tag{5.125}$$

Thus, we obtain the solution

$$x = a\left\{1 - \frac{1}{(1+(n-1)ka^{n-1}t)^{1/(n-1)}}\right\}, \tag{5.126}$$

which is the exact solution.

5.3.5. Variational Approach to Solitary Solution

In the review article [7], the variational approach to solitons was outlined by few lines, and now the method has been successfully applied to the search for soliton solutions [26, 27] without requirement of small parameter assumption, leading to an extremely simple and elementary but rigorous derivation of soliton solutions.

Considering the KdV equation, we seek its traveling wave solutions in the following frame:

$$u(x,t) = U(\xi), \quad v(x,t) = V(\xi), \quad \xi = x - ct, \tag{5.127}$$

where c is angular frequency. Substituting (5.127) into (4.1) yields

$$-cu' - 6uu' + u''' = 0, \tag{5.128}$$

where prime denotes the differential with respect to ξ.

Integrating (5.128) yields the result

$$-cu - 3u^2 + u'' = A, \tag{5.129a}$$

where A is an integration constant, which can be determined from the initial condition. For solitary solutions or limit cycles, the solutions do not depend upon the initial condition, so we always set $A = 0$, and this results in

$$-cu - 3u^2 + u'' = 0. \tag{5.129b}$$

By the semi-inverse method [17], the following variational formulation is established:

$$J = \int_0^\infty \left(\frac{1}{2}cu^2 + u^3 + \frac{1}{2}\left(\frac{du}{d\xi}\right)^2\right)d\xi. \tag{5.130}$$

Equation (5.130) also hints a Hamiltonian invariant, which reads

$$\frac{1}{2}cu^2 + u^3 - \frac{1}{2}u'^2 = B, \tag{5.131a}$$

where B is a constant, which can be determined by incorporating the initial condition. For solitary solutions or limit cycles, we set $B = 0$,

$$\frac{1}{2}cu^2 + u^3 - \frac{1}{2}u'^2 = 0. \tag{5.131b}$$

Differentiating (5.131a) with respect to ξ, we have

$$cuu' + 3u^2u' - u'u'' = 0, \tag{5.132}$$

which is equivalent to (5.129b).

The Hamiltonian invariant (5.131a) and (5.131b) can also be used for construction of an asymptotic solution or used as an auxiliary function in the subequation method, see Section 5.10.

The semi-inverse method is a powerful mathematical tool to the search for variational formulae for real-life physical problems.

By Ritz method, we search for a solitary wave solution in the form

$$u = p\,\mathrm{sech}^2(q\xi), \tag{5.133}$$

where p and q are constants to be further determined.

Substituting (5.133) into (5.130) results in

$$\begin{aligned} J &= \int_0^\infty \left[\frac{1}{2}cp^2\mathrm{sech}^4(q\xi) + p^3\mathrm{sech}^6(q\xi) + \frac{1}{2}(4p^2q^2\mathrm{sech}^4(q\xi)\tanh^2(q\xi)\right] d\xi \\ &= \frac{cp^2}{2q}\int_0^\infty \mathrm{sech}^4(z)dz + \frac{p^3}{q}\int_0^\infty \mathrm{sech}^6(z)dz + 2p^2q\int_0^\infty \left\{\mathrm{sech}^4(z)\tanh^4(z)\right\}dz \\ &= \frac{cp^2}{3q} + \frac{8p^3}{15q} + \frac{4p^2q}{15}. \end{aligned} \tag{5.134}$$

Making J stationary with respect to p and q results in

$$\begin{aligned} \frac{\partial J}{\partial p} &= \frac{2cp}{3q} + \frac{24p^2}{15q} + \frac{8pq}{15} = 0, \\ \frac{\partial J}{\partial q} &= -\frac{cp^2}{3q^2} - \frac{8p^3}{15q^2} + \frac{4p^2}{15} = 0, \end{aligned} \tag{5.135}$$

or simplifying

$$\begin{aligned} 5c + 12p + 4q^2 &= 0, \\ -5c - 8p + 4q^2 &= 0. \end{aligned} \tag{5.136}$$

From (5.136), we can easily obtain the following relations:

$$p = -\frac{1}{2}c, \qquad q = \sqrt{\frac{c}{4}}. \tag{5.137}$$

So the solitary wave solution can be approximated as

$$u = -\frac{c}{2}\operatorname{sech}^2\sqrt{\frac{c}{4}}(x - ct - \xi_0), \tag{5.138}$$

which is the exact solitary wave solution of KdV equation.

For the KdV equation expressed in (5.47), we have the following variational principle (see (5.54)):

$$J(u) = \iint_{-\infty}^{+\infty}\left\{-\frac{1}{2}u_x u_t + \frac{1}{2}u_{xx}^2 - \frac{1}{2}uu_x^2 - \frac{1}{4}u^2 u_{xx}\right\}dxdt. \tag{5.139}$$

We assume that the solitary solution can be expressed in the form

$$u(x,t) = \frac{1}{a_1 e^{a_5 x - a_5 t} + a_2 e^{-a_7 x + a_8 t} + a_3}, \tag{5.140}$$

where $a_1, a_2, \ldots, a_8$ are unknown constants to be further determined.

Substituting (5.140) into (5.139), we have

$$J(u) = J(a_1, a_2, \ldots, a_8). \tag{5.141}$$

Making the functional, (5.139), stationary can be approximated achieved by making the function (5.141) stationary, and this requires

$$\frac{\partial J}{\partial a_i} = 0 \quad (i = 1 \sim 8). \tag{5.142}$$

Solving the system, (5.142), we can determine the values for $a_1, a_2, \ldots, a_8$.

5.4. Hamiltonian Approach to Solitary Solution

5.4.1. Generalized Action Principles in Mechanics

We begin with the definition of the action functional as time integral over the Lagrangian L of a dynamical system

$$J(x_i) = \int_{t_1}^{t_2} L\,dt, \tag{5.143}$$

and the Lagrangian is defined as follows:

$$L = \frac{1}{2}mx_i'^2 - V(x_i). \tag{5.144}$$

Here, $x_i' = dx_i/dt$.

Newton's motion equation can be obtained from the stationary condition of the functional (5.143), which reads

$$mx_i'' + \frac{\partial V}{\partial x_i} = 0. \tag{5.145}$$

We can introduce some constraints to the action functional (5.143), leading to various principles required. For example, if the total energy is a conserved quantity, that is, $T + V =$ const., which is considered as a constraint of the functional (5.143), then we obtain the Euler-Maupertuis principle (principle of least action) [28]:

$$\int_{t_1}^{t_2} (2T - \text{const.})dt \longrightarrow \min, \tag{5.146}$$

or

$$\int_{t_1}^{t_2} T dt \longrightarrow \min. \tag{5.147}$$

In this section, we will obtain Hamiltonian and other actions from the Lagrangian (5.144) by introducing some constraints.

Now we introduce a generalized velocity

$$p_i = \frac{\partial L}{\partial x_i'} = mx_i'. \tag{5.148}$$

We consider (5.148) as a constraint of the action functional (5.143), and accordingly the Lagrangian (5.144) can be written as follows:

$$L(x_i, p_i) = \frac{1}{2m}p_i^2 - V(x_i). \tag{5.149}$$

By the Lagrange multiplier, we have the following generalized Lagrangian:

$$L_1(x_i, p_i, \lambda_i) = \frac{1}{2m}p_i^2 - V(x_i) + \lambda_i(p_i - mx_i'). \tag{5.150}$$

The multiplier can be readily identified, which reads

$$\lambda_i = -\frac{1}{m}p_i. \tag{5.151}$$

Substituting the identified multiplier into (5.150) results in

$$L_1(x_i, p_i) = \frac{1}{2m}p_i^2 - V(x_i) - \frac{1}{m}p_i(p_i - mx_i') = p_i x_i' - H, \tag{5.152}$$

where $H(x_i, p_i)$ is a Hamiltonian:

$$H(x_i, p_i) = \frac{1}{2m}p_i^2 + V(x_i). \tag{5.153}$$

If we introduce a new variable u_i defined as

$$u_i = x_i', \tag{5.154}$$

we consider (5.154) as a constraint of the action functional (5.143), and in such case, the Lagrangian (5.144) can be rewritten as

$$L(x_i, u_i) = \frac{1}{2}mu_i^2 - V(x_i). \tag{5.155}$$

By the Lagrange multiplier, we have the following generalized Lagrangian:

$$L_2(x_i, u_i, \lambda_i) = \frac{1}{2}mu_i^2 - V(x_i) + \lambda_i(u_i - x_i'). \tag{5.156}$$

The multiplier can be readily identified, which reads

$$\lambda_i = -mu_i. \tag{5.157}$$

Substituting the identified multiplier into (5.156) results in

$$L_2(x_i, u_i) = \frac{1}{2}mu_i^2 - V(x_i) - mu_i(u_i - x_i') = mu_i x_i' - \widetilde{H}(x_i, u_i), \tag{5.158}$$

where $\widetilde{H}(x_i, u_i)$ is given by

$$\widetilde{H}(x_i, p_i) = \frac{1}{2}mu_i^2 + V(x_i). \tag{5.159}$$

From (5.157), we know that the multiplier is actually the generalized velocity

$$\lambda_i = -mu_i = -p_i. \tag{5.160}$$

Substituting (5.160) into (5.156), and keeping p_i an independent variable, we have

$$L_3(x_i, u_i, p_i) = \frac{1}{2}mu_i^2 - V(x_i) - p_i(u_i - x_i') = p_i x_i' - \overline{H}(x_i, u_i, p_i), \tag{5.161}$$

where

$$\overline{H}(x_i, u_i, p_i) = -\frac{1}{2}mu_i^2 + V(x_i) + p_i u_i. \tag{5.162}$$

Equation (5.161) is called the Schwinger action [28].

By the same manipulation, from (5.151), the multiplier can be also determined as

$$\lambda_i = -\frac{1}{m}p_i = -u_i. \tag{5.163}$$

So we obtain another action like Schwinger's, which reads

$$L_4(x_i, p_i, u_i) = \frac{1}{2m}p_i^2 - V(x_i) - u_i(p_i - mx_i') = mu_i x_i' - \widehat{H}(x_i, u_i, p_i), \tag{5.164}$$

where

$$\widehat{H}(x_i, u_i, p_i) = -\frac{1}{2m}p_i^2 + V(x_i) + u_i p_i. \tag{5.165}$$

A more generalized action can be obtained by linear combination of $L_1(x_i, u_i)$ and $L_2(x_i, p_i)$:

$$\begin{aligned} L_5(x_i, u_i, p_i) &= L_1(x_i, u_i) + L_2(x_i, p_i) \\ &= -\frac{1}{2m}p_i^2 - \frac{1}{2}mu_i^2 + (p_i + mu_i)x_i' - 2V(x_i) \\ &= (p_i + mu_i)x_i' - H_e(x_i, u_i, p_i), \end{aligned} \tag{5.166}$$

where

$$H_e(x_i, u_i, p_i) = \frac{1}{2m}p_i^2 + \frac{1}{2}mu_i^2 + 2V(x_i). \tag{5.167}$$

The Euler equations can be readily obtained, which read

$$\begin{aligned}
\delta x_i &: \frac{d}{dt}(p_i + mu_i) + \frac{\partial H_e}{\partial x_i} = \frac{d}{dt}(p_i + mu_i) + 2\frac{\partial V}{\partial x_i} = 0, \\
\delta u_i &: mx_i' - \frac{\partial H_e}{\partial u_i} = mx_i' - mu_i = 0, \\
\delta p_i &: x_i' - \frac{\partial H_e}{\partial p_i} = x_i' - \frac{1}{m}p_i = 0.
\end{aligned} \tag{5.168}$$

In a more general form, (5.166) can be written as

$$L_5(x_i, u_i, p_i) = \alpha L_1(x_i, u_i) + \beta L_2(x_i, p_i), \tag{5.169}$$

where α and β are constants.

Linearly combining $L_i (i = 1, 2, 3, 4, 5)$, we have

$$L_6(x_i, u_i, p_i) = \sum_{i=1}^{5} \alpha_i L_i, \tag{5.170}$$

where α_i are constants, and we often set $\sum_{i=1}^{5} \alpha_i = 1$.

5.4.2. Modified Hamilton Principles for Initial Value Problems

The Hamilton principle can be written in the form

$$J(u_i) = \int_0^T \left(\frac{1}{2} m\dot{u}_i^2 - U \right) dt, \tag{5.171}$$

with the special requirements $\delta u(0) = \delta u(T) = 0$.

Here u_i is the velocity component in x_i, and U is a potential defined as $\partial U/\partial x_i = -F_i$, where F_i is the body force component in x_i.

The Hamilton's principle holds only for the conditions prescribed at the beginning and at the end of the motion and is therefore useless to deal with the usual initial condition problems, both as an analytical tool and as a basis for approximate solution methods. It is impossible for most real-life physical problems to prescribe terminal conditions.

In order to eliminate the unnecessary final condition at $t = T$, Carini and Genna [29] obtained the following functional for the case $u_i(0) = \dot{u}_i(0) = 0$:

$$J(u_i) = \int_0^T \left(\frac{1}{2} m\ddot{u}_i^2 - \ddot{F}_i u_i \right) dt + \dot{F}_i(T)u_i(T) - F_i(T)\dot{u}_i(T). \tag{5.172}$$

We found that the natural final conditions at $t = T$ satisfy the physical requirements.

Liu [30] considered that, in order to deal with the final condition, one term should be added to the Hamilton principle to result in the following functional:

$$J(u_i) = \int_0^T \left(\frac{1}{2}m\dot{u}_i^2 - U\right)dt - m\tilde{\dot{u}}_i\dot{u}_i\Big|_{t=T}, \tag{5.173}$$

where $\tilde{\dot{u}}_i$ is a restricted variable, that is, $\delta\tilde{\dot{u}}_i = 0$. The modification, (5.172), is not a classic variational principle due to that the variable $\tilde{\dot{u}}_i$ has to be prescribed at $t = T$, so that

$$\delta\dot{u}_i(T) = 0, \tag{5.174}$$

and this requirement is even more overrestricted than $\delta u_i(T) = 0$ in Hamilton principle, so no vital innovation was made due to its inconsistent at the final condition. However, Liu [30] obtained some functionals that can successfully deal with the initial conditions, but the final condition still keeps an issue of polemics.

It would be a landmark in the history of calculus of variations after Hamilton if we can extend the principle to all initial-value problems without prescribing both initial and final conditions. In order to deal with the final condition, we consider the conserved Hamiltonian:

$$\frac{1}{2}m\sum_{i=1}^{3}\dot{u}_i^2 + U = \frac{1}{2}m\sum_{i=1}^{3}\dot{u}_{i0}^2 + U_0. \tag{5.175}$$

Here $U_0 = U|_{t=0}$, so we have the following identity:

$$\dot{u}_i = \sqrt{\sum_{i=1}^{3}\dot{u}_{i0}^2 - \sum_{j=1}^{3}\dot{u}_j^2 + \frac{2(U_0 - U)}{m}}, \quad i \neq j, \tag{5.176}$$

for all $t \geq 0$.

In order to convert the initial conditions $u_i(0) = u_{i0}$ and $\dot{u}_i(0) = \dot{u}_{i0}$ into natural initial conditions and make the natural final condition satisfy the physical requirement, (5.176), we assume that its modified Hamilton principle can be written in the form

$$J(u_i) = \int_0^T \left(\frac{1}{2}m\dot{u}_i^2 - U\right)dt + f(u_i, \dot{u}_i)|_{t=0}^{t=T}, \tag{5.177}$$

where f is an unknown function of u_i and/or its derivatives.

Making the above functional stationary, we obtain the following stationary conditions at $t = 0$ and $t = T$:

$$\begin{aligned} \left[m\dot{u}_i + \frac{\partial f}{\partial u_i}\right]_{t=0}^{t=T} &= 0, \\ \frac{\partial f}{\partial \dot{u}_i}\Big|_{t=0}^{t=T} &= 0. \end{aligned} \tag{5.178}$$

We search such an f so that the above 4 equations satisfy all initial conditions at $t = 0$ and meet exactly the physical requirement, for example, (5.176), at $t = T$.

Accordingly, we can assume that

$$\left(\frac{\partial f}{\partial u_i}\right)_{t=0} = m\dot{u}_{i0}|_{t=0},$$
$$\left(\frac{\partial f}{\partial \dot{u}_i}\right)_{t=0} = k(u_i - u_{i0})|_{t=0}, \tag{5.179}$$
$$\left(\frac{\partial f}{\partial u_i}\right)_{t=T} = -m\dot{u}_i|_{t=T} = -m\sqrt{\sum_{i=1}^{3}\dot{u}_{i0}^2 - \sum_{j=1}^{3}\dot{u}_j^2 + \frac{2(U_0 - U)}{m}}\Bigg|_{t=T}, \quad i \neq j.$$

From the above relations, we can identify f in the form

$$f = \left\{m\dot{u}_{i0}u_i + k\left(\frac{1}{2}\dot{u}_i^2 - \dot{u}_{i0}\dot{u}_i\right)\right\}_{t=0} + \mathrm{He}|_{t=T}, \tag{5.180}$$

where k is a nonzero constant; He is defined as

$$\mathrm{He} = \frac{m^2}{3F_i}\left[\sum_{i=1}^{3}\dot{u}_{i0}^2 - \sum_{j=1}^{3}\dot{u}_j^2 + \frac{2(U_0 - U)}{m}\right]^{3/2}. \tag{5.181}$$

So we obtain the following modified Hamilton principle:

$$J(u_i) = \int_0^T\left(\frac{1}{2}m\dot{u}_i^2 - U\right)dt + \left\{m\dot{u}_{i0}u_i + k\left(\frac{1}{2}\dot{u}_i^2 - \dot{u}_{i0}\dot{u}_i\right)\right\}_{t=0} + \mathrm{He}|_{t=T}, \tag{5.182}$$

which holds for all initial value problems.

Proof. Making the obtained functional (5.182) stationary, we obtain

(1) in the solution domain ($0 < t < T$):

$$-m\ddot{u}_i + F_i = 0, \tag{5.183}$$

which is Newton's motion equation;

(2) natural initial conditions ($t = 0$):

$$-m\dot{u}_i|_{t=0} + m\dot{u}_{i0}|_{t=0} = 0,$$
$$\{k(\dot{u}_i - \dot{u}_{i0})\}_{t=0} = 0, \tag{5.184}$$

which satisfy obviously the initial conditions $u_i(0) = u_{i0}$ and $\dot{u}_i(0) = \dot{u}_{i0}$, respectively;

(3) natural final condition ($t = T$):

$$m\dot{u}_i|_{t=T} - m\left.\sqrt{\sum_{i=1}^{3}\dot{u}_{i0}^2 - \sum_{j=1}^{3}\dot{u}_j^2 + \frac{2(U_0 - U)}{m}}\right|_{t=T} = 0, \tag{5.185}$$

which meet the physical requirement of (5.176) at $t = T$.

We have alternative approaches to identifying f by different assumptions in (5.178), leading to various new modifications. We write here few modified Hamilton principles for reference

$$J(u_i) = \int_0^T\left(\frac{1}{2}m\dot{u}_i^2 - U\right)dt + \left\{m\dot{u}_{i0}u_i + k_1\left(\frac{1}{2}u_i^2 - u_{i0}u_i\right) + k_2\left(\frac{1}{2}\dot{u}_i^2 - \dot{u}_{i0}\dot{u}_i\right)\right\}_{t=0} + \mathrm{He}|_{t=T},$$

$$J(u_i) = \int_0^T\left(\frac{1}{2}m\dot{u}_i^2 - U\right)dt$$
$$+ \{(m - k_4)u_{i0}u_i - k_3u_{i0}\dot{u}_i + k_3u_i(\dot{u}_i - \dot{u}_{i0}) + k_4\dot{u}_i(u_i - u_{i0})\}_{t=0} + \mathrm{He}|_{t=T}, \tag{5.186}$$

where k's are nonzero constants, and it requires that $k_1 - m \neq 0$ and $k_3 + k_4 = 0$.

To summarize, we can conclude from the above derivation and strict proof that the obtained modified Hamilton principles, which are first deduced in the history, and valid for all initial-value problems, are extremely important in both pure and applied sciences due to complete elimination of the long-existing shortcomings in Hamilton principle. The stationary conditions of the obtained variational principle satisfy the Newton's motion equation, and all initial conditions, furthermore, the natural final condition ($t = T$), satisfy automatically the physical requirement, making a vital innovation of Hamilton principle. □

5.4.3. Hamiltonian Approach to Nonlinear Oscillators [31, 32]

In this paper, we consider the following general oscillator:

$$u'' + f(u) = 0, \tag{5.187}$$

with initial conditions $u(0) = A$ and $u'(0) = 0$.

It is easy to establish a variational principle for (5.187), which reads [31]

$$J(u) = \int_0^{T/4}\left\{\frac{1}{2}u'^2 - F(u)\right\}dt, \tag{5.188}$$

where T is the period of the oscillator, $\partial F/\partial u = f(u)$.

In the functional (5.188), $(1/2)u'^2$ is kinetic energy, and $F(u)$ is potential energy, so the functional (5.188) is the least Lagrangian action, from which we can immediately obtain its Hamiltonian, which reads

$$H = \frac{1}{2}u'^2 + F(u) = \text{constant} = H_0, \tag{5.189}$$

or

$$\frac{1}{2}u'^2 + F(u) - H_0 = 0. \tag{5.190}$$

Equation (5.189) replies that the total energy keeps unchanged during the oscillation.

Assume that the solution can be expressed as

$$u = A\cos\omega t, \tag{5.191}$$

where ω is the frequency.

Submitting (5.191) into (5.190) results in a residual

$$R(t) = \frac{1}{2}A^2\omega^2\sin^2\omega t + F(A\cos\omega t) - H_0. \tag{5.192}$$

According to the energy balance method proposed in [33], locating at some a special point, that is, $\omega t = \pi/4$, and setting $R(t = \pi/4\omega) = 0$, we can obtain an approximate frequency-amplitude relationship of the studied nonlinear oscillator. Such treatment is much simple and has been widely used by engineers [34–36]. The accuracy of such location method, however, strongly depends upon the chosen location point. To overcome the shortcoming of the energy balance method, in this paper, we suggest a new approach based on Hamiltonian.

From (5.192), we have

$$\frac{\partial H}{\partial A} = 0. \tag{5.193}$$

Introducing a new function, $\overline{H}(u)$, defined as

$$\overline{H}(u) = \int_0^{T/4} \left\{\frac{1}{2}u'^2 + F(u)\right\} dt = \frac{1}{4}TH, \tag{5.194}$$

it is obvious that

$$\frac{\partial \overline{H}}{\partial T} = \frac{1}{4}H. \tag{5.195}$$

Equation (5.195) is, then, equivalent to the following one:

$$\frac{\partial}{\partial A}\left(\frac{\partial \overline{H}}{\partial T}\right) = 0, \tag{5.196}$$

or

$$\frac{\partial}{\partial A}\left(\frac{\partial \overline{H}}{\partial (1/\omega)}\right) = 0. \tag{5.197}$$

From (5.197), we can obtain approximate frequency-amplitude relationship of a nonlinear oscillator.

Consider the Duffing equation

$$u'' + u + \varepsilon u^3 = 0, \qquad u(0) = A, \qquad u'(0) = 0. \tag{5.198}$$

Its Hamiltonian can be easily obtained, which reads

$$H = \frac{1}{2}u'^2 + \frac{1}{2}u^2 + \frac{1}{4}\varepsilon u^4. \tag{5.199}$$

Integrating (5.199) with respect to t from 0 to $T/4$, we have

$$\overline{H}(u) = \int_0^{T/4}\left\{\frac{1}{2}u'^2 + \frac{1}{2}u^2 + \frac{1}{4}\varepsilon u^4\right\}dt. \tag{5.200}$$

Assuming that the solution can be expressed as $u = A\cos\omega t$ and substituting it to (5.200), we obtain

$$\begin{aligned}\overline{H}(u) &= \int_0^{T/4}\left\{\frac{1}{2}A^2\omega^2\sin^2\omega t + \frac{1}{2}A^2\cos^2\omega t + \frac{1}{4}\varepsilon A^4\cos^4\omega t\right\}dt \\ &= \int_0^{\pi/2}\left\{\frac{1}{2}A^2\omega\sin^2 t + \frac{1}{\omega}\left[\frac{1}{2}A^2\cos^2 t + \frac{1}{4}\varepsilon A^4\cos^4 t\right]\right\}dt \\ &= \frac{1}{2}A^2\omega\cdot\frac{\pi}{4} + \frac{1}{\omega}\left[\frac{1}{2}A^2\cdot\frac{\pi}{4} + \frac{1}{4}\varepsilon A^4\cdot\frac{3}{4}\cdot\frac{\pi}{4}\right].\end{aligned} \tag{5.201}$$

Setting

$$\frac{\partial}{\partial A}\left(\frac{\partial \overline{H}}{\partial (1/\omega)}\right) = -A\omega^2\cdot\frac{\pi}{4} + \left[A\cdot\frac{\pi}{4} + \varepsilon A^3\cdot\frac{3}{4}\cdot\frac{\pi}{4}\right] = 0, \tag{5.202}$$

we obtain the following frequency-amplitude relationship:

$$\omega = \sqrt{1 + \frac{3}{4}\varepsilon A^3}. \tag{5.203}$$

Now we consider another nonlinear oscillator with discontinuity:

$$\frac{d^2u}{dt^2} + \operatorname{sgn}(u) = 0, \qquad u(0) = A, \qquad u'(0) = 0. \tag{5.204}$$

$\operatorname{sgn}(u)$ is +1 and −1 for $u > 0$ and $u < 0$, respectively.

Its variational formulation can be written as

$$J(u) = \int_0^{T/4} \left(-\frac{1}{2}u'^2 + u\right) dt + \int_{T/4}^{T/2} \left(-\frac{1}{2}u'^2 - u\right) dt, \tag{5.205}$$

and $\overline{H}(u)$ can be written in the form

$$\overline{H}(u) = \int_0^{T/4} \left(\frac{1}{2}u'^2 + u\right) dt + \int_{T/4}^{T/2} \left(\frac{1}{2}u'^2 - u\right) dt. \tag{5.206}$$

Using $u = A\cos\omega t$ as an approximate solution, we have

$$\begin{aligned}
\overline{H} &= \int_0^{T/4} \left\{\frac{1}{2}A^2\omega^2\sin^2\omega t + A\cos\omega t\right\} dt + \int_{T/4}^{T/2} \left\{\frac{1}{2}A^2\omega^2\sin^2\omega t - A\cos\omega t\right\} dt \\
&= \int_0^{\pi/2} \left\{\frac{1}{2}A^2\omega\sin^2 t + \frac{1}{\omega}A\cos t\right\} dt + \int_{\pi/2}^{\pi} \left\{\frac{1}{2}A^2\omega\sin^2 t - \frac{1}{\omega}A\cos t\right\} dt \\
&= \frac{1}{2}A^2\omega \cdot \frac{\pi}{4} + \frac{1}{\omega}A + \frac{1}{2}A^2\omega \cdot \frac{\pi}{4} + \frac{1}{\omega}A = A^2\omega \cdot \frac{\pi}{4} + \frac{2}{\omega}A.
\end{aligned} \tag{5.207}$$

Setting

$$\frac{\partial}{\partial A}\left(\frac{\partial \overline{H}}{\partial(1/\omega)}\right) = -A\omega^2 \cdot \frac{\pi}{2} + 2 = 0, \tag{5.208}$$

we obtain

$$\omega = \frac{2}{\sqrt{\pi A}} = \frac{1.128379}{\sqrt{A}}, \tag{5.209}$$

which is very close to the exact one

$$\omega_{\text{exact}}(A) = \frac{\pi}{2\sqrt{2A}} = \frac{1.110721}{\sqrt{A}}. \tag{5.210}$$

The 1.59% accuracy is acceptable considering the simple solution procedure.

5.4.4. Hamilton Approach to Solitary Solutions

The Hamiltonian approach to nonlinear oscillators has been now widely used [37–40]; in this section, we will extend the technology to nonlinear wave equations.

The Euler-Lagrange equation of (5.130) reads

$$\frac{dL}{du} - \frac{d}{d\xi}\left(\frac{dL}{du_\xi}\right) = 0, \tag{5.211}$$

where $L = (1/2)cu^2 + u^3 + (1/2)(du/d\xi)^2$ and $u_\xi = du/d\xi$.

It is obvious that the above Euler-Lagrange equation is the KdV equation.

We write the Lagrange function in the form $L = T - V$, where T is the kinetic energy defined as

$$T = \frac{1}{2}\left(\frac{du}{d\xi}\right)^2, \tag{5.212}$$

and V is potential energy defined as

$$V = -\frac{1}{2}cu^2 - u^3. \tag{5.213}$$

The Hamiltonian invariant reads $T + V = B$, that is,

$$\frac{1}{2}u'^2 - \frac{1}{2}cu^2 - u^3 = B, \tag{5.214}$$

where B is a constant, which can be determined by incorporating the initial condition. For solitary solutions or limit cycles, we set $B = 0$. Equation (5.214) becomes

$$\frac{1}{2}cu^2 + u^3 - \frac{1}{2}u'^2 = 0. \tag{5.215}$$

Differentiating (5.215) with respect to ξ, we have

$$cuu' + 3u^2u' - u'u'' = 0, \tag{5.216}$$

which is equivalent to (5.129b) in case $u' \neq 0$.

If a solitary solution is solved, considering tail property of a soliton, we assume that

$$u(\xi) = \frac{\sum_n a_n \exp(q_n \xi)}{\sum_m b_m \exp(p_m \xi)}, \tag{5.217}$$

where a_i, b_i, p_i, and q_i are constants to be further determined asymptotically. Equation (5.217) should satisfy (2.2).

The assumption, (5.217), is also used in the exp-function method (see Section 5.9), where the unknowns are identified exactly, whereas we will determine their values approximated by the weighted residual method.

Substituting (5.217) into (5.216) results in the following residual:

$$R(\xi) = \frac{1}{2}cu^2 + u^3 - \frac{1}{2}u'^2. \tag{5.218}$$

We can use the method of the least squares to determine the unknown constants involved in (5.217)

$$\int_{-\infty}^{+\infty} R^2(\xi) d\xi \longrightarrow \min. \tag{5.219}$$

We can also use the location method to simply identify the unknowns. To elucidate the solution procedure, we assume its solitary solution has a simple symmetrical form

$$u(\xi) = \frac{1}{a(e^{k\xi} + e^{-k\xi}) + b}, \tag{5.220}$$

where a, b, and k are unknown constants.

The residual equation is

$$R(\xi) = \frac{c}{2\left[a(e^{k\xi} + e^{-k\xi}) + b\right]^2} + \frac{1}{\left[a(e^{k\xi} + e^{-k\xi}) + b\right]^3} - \frac{\left[ak(e^{k\xi} - e^{-k\xi})\right]^2}{\left[a(e^{k\xi} + e^{-k\xi}) + b\right]^4}. \tag{5.221}$$

To determine the values of the unknowns in (5.220), we set

$$\begin{aligned} R(0) &= 0, \\ R''(0) &= 0, \\ R^{(4)}(0) &= 0. \end{aligned} \tag{5.222}$$

Note that $R'(0) = R'''(0) \equiv 0$.

Simplifying (5.222), we obtain

$$\begin{aligned}
2ac + bc + 2 &= 0, \\
2ac + bc + 3 + 2ak^2 &= 0, \\
32a^2c + 14abc - b^2c + 66a + 80a^2bk^2 - 3b &= 0.
\end{aligned} \tag{5.223}$$

Solving (5.223) simultaneously, we have

$$\begin{aligned}
a &= -\frac{1}{2k^2}, \\
b &= -\frac{1}{k^2}, \\
c &= k^2.
\end{aligned} \tag{5.224}$$

We, therefore, obtain the following approximate solitary solution:

$$\begin{aligned}
u(x,t) &= \frac{1}{-(1/2k^2)\{\exp(k(x-k^2t) + \exp(-k(x-k^2t)))\} - 1/k^2} \\
&= \frac{-2k^2}{\{\exp k(x-k^2t) + \exp(-k(x-k^2t))\} + 2}.
\end{aligned} \tag{5.225}$$

Submitting the obtained solution into (4.1), we find it as an exact solution!

Consider the following generalized KdV equation:

$$u_t + 6u^n u_x + u_{xxx} = 0. \tag{5.226}$$

This equation admits no variational formulation. If a traveling wave solution is searched for, we can make a transformation $\xi = x - ct$, and then (5.226) reduces to an ordinary differential equation

$$-cu' + 6u^n u' + u''' = 0. \tag{5.227}$$

Integrating the above equation results in

$$-cu + \frac{6}{n+1}u^{n+1} + u'' = A, \tag{5.228}$$

which admits a variational formulation

$$J(u) = \int_{-\infty}^{+\infty} \left\{ -\frac{1}{2}cu^2 + \frac{6}{(n+1)(n+2)}u^{n+2} - Au - \frac{1}{2}u'^2 \right\} d\eta. \tag{5.229}$$

We write $T = (1/2)u'^2$ and $E = -(1/2)cu^2 + (6/(n+1)(n+2))u^{n+2} - Au$, and then T can be interpreted as "kinetic energy" and E "potential energy." The variational functional, (5.229), can be rewritten in the form of Lagrangian action

$$J(u) = -\int_{-\infty}^{+\infty} \{T - E\} d\eta, \tag{5.230}$$

and the Hamiltonian invariant implies the conversation of energy, which requires

$$-\frac{1}{2}cu^2 + \frac{6}{(n+1)(n+2)}u^{n+2} - Au + \frac{1}{2}u'^2 = B, \tag{5.231}$$

or

$$u'^2 = cu^2 - \frac{12}{(n+1)(n+2)}u^{n+2} + 2Au + 2B. \tag{5.232}$$

For solitary solutions, $A = B = 0$, and (5.232) reduces to

$$u'^2 = cu^2 - \frac{12}{(n+1)(n+2)}u^{n+2}. \tag{5.233}$$

If a solitary solution is solved, we assume that

$$u(\xi) = \frac{1}{ae^{k\xi} + be^{-k\xi} + d}. \tag{5.234}$$

Substituting (5.234) into (5.233) results in the following residual:

$$R(\xi) = u'^2 - cu^2 + \frac{12}{(n+1)(n+2)}u^{n+2}. \tag{5.235}$$

Proceeding a similar way as that for (5.222), we can identify a, b, and d in (5.234), and finally, we obtain

$$u(x,t) = \frac{-2k^2}{\exp(kx - k^3t + D) + \exp[-(kx - k^3t + D)] - 2}. \tag{5.236}$$

For a compacton solution, we assume that

$$u(\xi) = \frac{a\cos^2\xi}{b + c\cos^2\xi}. \tag{5.237}$$

By a similar solution procedure as above, we can determine a, b, and c with ease.

5.5. Variational Principles for Fluids

5.5.1. A Bernoulli-Like Equation for Shallow Wave Propagation

Strong storms and cyclones, underwater earthquakes, high-speed ferries, and aerial and submarine landslides can cause giant surface waves approaching the coast and frequently cause extensive coastal flooding, destruction of coastal constructions, and loss of lives. A fast but reliable prediction of a tsunami pulse is of critical importance, and a simple equation like Bernoulli equation is, therefore, much needed.

To this end, we use a one-dimensional nonlinear shallow wave propagation for describing runup of irregular waves on a beach. The basic equations are [41]

$$\frac{\partial u}{\partial t}+u\frac{\partial u}{\partial x}+g\frac{\partial \eta}{\partial x}=0, \tag{5.238}$$

$$\frac{\partial \eta}{\partial t}+\frac{\partial}{\partial x}[(h+\eta)u]=0, \tag{5.239}$$

where η is water displacement, u is the depth-averaged velocity, h is the unperturbed water depth, and g is the gravity acceleration.

We will establish a variational principle for the system of (5.238) and (5.239) to search for a principle of conservation of energy for the present problem.

In order to establish a variational formulation for the system of (5.238) and (5.239), we introduce a function Φ defined as

$$\begin{gathered}\frac{\partial \Phi}{\partial x}=u,\\ \frac{\partial \Phi}{\partial t}=-\left(\frac{1}{2}u^2+g\eta\right).\end{gathered} \tag{5.240}$$

Equation (5.240) is equivalent to (5.238).

Using the semi-inverse method [17], we construct a trial functional in the form

$$J(\Phi,u,\eta)=\iint\left\{\eta\frac{\partial \Phi}{\partial t}+(h+\eta)u\frac{\partial \Phi}{\partial x}+F(u,\eta)\right\}dt\,dx, \tag{5.241}$$

where F is an unknown function of u and η.

It is obvious that the stationary condition of (5.241) with respect to Φ is (5.239). Now making the functional, (5.241), stationary with respect to η and u, we obtain the following Euler-Lagrange equations:

$$\begin{gathered}\frac{\partial \Phi}{\partial t}+u\frac{\partial \Phi}{\partial x}+\frac{\partial F}{\partial \eta}=0,\\ (h+\eta)\frac{\partial \Phi}{\partial x}+\frac{\partial F}{\partial u}=0.\end{gathered} \tag{5.242}$$

The above equations should satisfy (5.240). To this end, using (5.240), we set

$$\frac{\partial F}{\partial \eta} = -\frac{\partial \Phi}{\partial t} - u\frac{\partial \Phi}{\partial x} = \frac{1}{2}u^2 + g\eta - u^2 = -\frac{1}{2}u^2 + g\eta,$$
$$\frac{\partial F}{\partial u} = -(h+\eta)\frac{\partial \Phi}{\partial x} = -(h+\eta)u. \tag{5.243}$$

From (5.243), the unknown F can be identified, which reads

$$F = -\frac{1}{2}u^2\eta + \frac{1}{2}g\eta^2 - \frac{1}{2}hu^2. \tag{5.244}$$

We, therefore, obtain the needed variational formulation, which is

$$J(\Phi, u, \eta) = \iint\left\{\eta\frac{\partial \Phi}{\partial t} + (h+\eta)u\frac{\partial \Phi}{\partial x} - \frac{1}{2}(\eta + h)u^2 + \frac{1}{2}g\eta^2\right\}dt\,dx. \tag{5.245}$$

Submitting (5.240) into (5.245), we obtain a constrained variational formulation

$$J(\Phi) = \iint\left\{\frac{1}{2}hu^2 - \frac{1}{2}g\eta^2\right\}dt\,dx, \tag{5.246}$$

which is under constraints of (5.240).

Proof. Making the functional, (5.246), stationary, we have

$$\delta J(\Phi) = \iint\{hu\delta u - g\eta\delta\eta\}dt\,dx. \tag{5.247}$$

According to the constraints, (5.240), we have

$$\delta\left(\frac{\partial \Phi}{\partial x}\right) = \delta u, \tag{5.248}$$

$$\delta\left(\frac{\partial \Phi}{\partial t}\right) = -\delta\left(\frac{1}{2}u^2 + g\eta\right) = -(u\delta u + g\delta\eta). \tag{5.249}$$

Replacing δu in (5.249) by $\delta(\partial\Phi/\partial x)$ results in

$$g\eta\delta\eta = -\eta\delta\left(\frac{\partial \Phi}{\partial t}\right) - \eta u\delta\left(\frac{\partial \Phi}{\partial x}\right). \tag{5.250}$$

Submitting (5.248) and (5.250) into (5.247), we obtain

$$\begin{aligned}
\delta J(\Phi) &= \iint \left\{ hu\delta\left(\frac{\partial\Phi}{\partial x}\right) + \eta\delta\left(\frac{\partial\Phi}{\partial t}\right) + \eta u\delta\left(\frac{\partial\Phi}{\partial x}\right) \right\} dt\, dx \\
&= \iint \left\{ (h+\eta)u\frac{\partial}{\partial x}(\delta\Phi) + \eta\frac{\partial}{\partial t}(\delta\Phi) \right\} dt\, dx \\
&= -\iint \left\{ \frac{\partial((h+\eta)u)}{\partial x} + \frac{\partial\eta}{\partial t} \right\} \delta\Phi dt\, dx + \iint \left\{ \frac{\partial}{\partial x}[(h+\eta)u\delta\Phi] + \frac{\partial}{\partial t}(\eta\delta\Phi) \right\} dt\, dx \\
&= -\iint \left\{ \frac{\partial((h+\eta)u)}{\partial x} + \frac{\partial\eta}{\partial t} \right\} \delta\Phi dt\, dx + \int_t (h+\eta)u\delta\Phi dt + \int_x \eta\delta\Phi dx.
\end{aligned} \tag{5.251}$$

Setting $\delta J(\Phi) = 0$, and ignoring the boundary terms, for arbitrary $\delta\Phi$, we can obtain (5.239) as the needed stationary condition.

We can rewrite (5.246) in the form

$$J(\Phi) = \iint \{T - V\} dt\, dx, \tag{5.252}$$

where $T = (1/2)hu^2$ is a generalized kinetic energy of water wave, is $V = (1/2)g\eta^2$ and a generalized gravitational potential.

Equation (5.252) is a principle of the Lagrange action for the shallow wave propagation, and it also implies conservation of energy $T + V = H$:

$$\frac{1}{2}hu^2 + \frac{1}{2}g\eta^2 = H, \tag{5.253}$$

where H is a constant.

We call (5.253) a Bernoulli-like equation for shallow wave propagation. □

Tsunami prediction, for example, after an earthquake, is of critical importance to save life and property. The extent of catastrophe depends mainly on the height of the Tsunami pulses, that is, the value of η. It is easy to observe the velocity of a wave in the sea, and we, therefore, can make easy but reliable tsunami prediction using the obtained Bernoulli-like equation.

5.5.2. Lin's Variational Principle for Ideal Flows

It is well known that the Hamilton's principle can be applied to a single fluid particle or a closed system by the involutory transformation. For an isoentropy rotational flow, a variational principle can be established using Lin's constraints [42]. But the essence of Lin's constraints is not clear yet, and this paper concludes that the functional with Lin's constraints is not a genuine variational principle, but an approximate one. In addition, a new generalized variational principle with only 6 independent variables (including only one Lin's constraint) for three-dimensional unsteady compressible rotational flow is established.

In 1950s, great progress had been made on the research of variational principle in solid mechanics, with Hu-Washizu variational principle [43] as milestone, and in fluid mechanics, with Lin's constraints [42] as milestone.

Hamilton's principle was so successfully and powerfully applied to particle mechanics that many attempts have been made to obtain the momentum equations from a variational principle patterned after Hamilton's principle. These attempts have not all been successful except in the case of isoentropy irrotational flow. For a more general one, Lin's constraints [44] must be added.

Considering a 3D unsteady inviscid compressible rotational flow, we have the following equations [45].

(1) Momentum equation:

$$\frac{\partial u_i}{\partial t} + u_i \frac{\partial u_i}{\partial x_i} = -\frac{1}{\rho}\frac{\partial P}{\partial x_i}, \tag{5.254}$$

where u_i is the flow velocity in x_i indirect, P is the pressure, and ρ is the density.

(2) Equation of state:

$$P = \rho^{\kappa} \exp\left\{\frac{S - S_0}{m}\right\}, \tag{5.255}$$

where κ is a specific heat ratio, $m = (\kappa - 1)^{-1}$, and S is entropy.

(3) Continuity equation:

$$\frac{\partial \rho}{\partial t} + \frac{\partial(\rho u_j)}{\partial x_j} = 0. \tag{5.256}$$

(4) Isoentropic equation:

$$\frac{\partial S}{\partial t} + u_j \frac{\partial S}{\partial x_j} = 0. \tag{5.257}$$

(5) Steady Bernoulli's equation:

$$\frac{1}{2}v^2 + \Pi = B. \tag{5.258}$$

For unsteady flow, we should use the following momentum equation:

$$\frac{\partial \vec{v}}{\partial t} + \nabla\left(\frac{1}{2}v^2\right) - \vec{v} \times (\nabla \times \vec{v}) + \nabla\Pi = 0, \tag{5.259}$$

where $\Pi = \int(dP/\rho)$ and $\vec{v} = u_1\vec{i} + u_2\vec{j} + u_3\vec{k}$.

From the equation of state, we can deduce the following equation:

$$\Pi = \frac{\kappa}{\kappa - 1}\rho^{\kappa-1}\exp\left\{\frac{S - S_0}{m}\right\}. \tag{5.260}$$

In 1955, Herivel first applied Hamilton's principle to fluid mechanics and deduced the following functional [46]:

$$J(x_i) = \int_t \int_{CV} \left\{\frac{1}{2}\rho u_i^2 - \rho E(\rho, S)\right\} dV dt, \tag{5.261}$$

where E is specific internal energy. CV is the control volume in Eulerian open space. Hamiltonian fluid mechanics becomes a branch of fluid mechanics and applied mathematics as well [47–56].

Applying the Lagrange multipliers to remove the constraint equations (5.257) and (5.256) yields

$$J = \int_t \int_{CV} \left\{\frac{1}{2}\rho u_i^2 - \rho E(\rho, S) - \rho\eta\left[\frac{\partial S}{\partial t} + u_j\frac{\partial S}{\partial x_j}\right] + \Phi\left[\frac{\partial \rho}{\partial t} + \frac{\partial(\rho u_j)}{\partial x_j}\right]\right\} dV dt. \tag{5.262}$$

Making the above functional stationary with respect to u_i, we have

$$\vec{v} = \nabla\Phi + \eta\nabla S. \tag{5.263}$$

However, according to the Crocco equation [45],

$$(\nabla \times \vec{v}) \times \vec{v} = T\nabla S - \nabla i^{\bullet}, \tag{5.264}$$

where $i^{\bullet}$ is enthalpy, T is temperature, and the homentropic flow may be a rotational one. Equation (5.263) implies, however, that homentropic flow ($\nabla S = 0$) must be a potential flow or irrotational flow ($\vec{v} = \nabla\Phi$), confiliting with (5.264), and therefore, the functional, equation (5.261) or (5.262), is not reasonable, which was just pointed out by Lin [44].

According to Clebsch [57], an arbitrary velocity vector can be expressed as follows:

$$\vec{v} = \nabla\Phi + \lambda_i + \nabla\tau_i, \tag{5.265}$$

where Φ, λ_i, and τ_i are Clebsch variables.

Lin introduced three-additional constraint equations (Lin's constraint equations) to the functional (5.262) [44]:

$$\frac{D\alpha_i}{Dt} = 0 \quad (i = 1, 2, 3), \tag{5.266}$$

where α_i is Lagrange variables in Lagrange space.

Introducing the Lagrange multipliers to remove Lin's constraints, we have the following general functional:

$$J=\int_t \int_V \left\{ \frac{1}{2}\rho u_i^2 - \rho E(\rho, S) - \rho\eta\left[\frac{\partial S}{\partial t} + u_j \frac{\partial S}{\partial x_j}\right] + \Phi\left[\frac{\partial \rho}{\partial t} + \frac{\partial(\rho u_j)}{\partial x_j}\right] - \rho\beta_i\left[\frac{\partial \alpha_i}{\partial t} + u_j\frac{\partial \alpha_i}{\partial x_j}\right]\right\} dV dt. \tag{5.267}$$

Making the above functional (5.267) stationary with respect to u_i, we have

$$\vec{v} = \nabla\Phi + \eta\nabla S + \beta_i \nabla\alpha_i. \tag{5.268}$$

This expression says that a homentropic flow ($\nabla S = 0$) can also be an rotational flow, agreeing with (5.264). So the functional (5.267) with Lin's constraints is acceptable for practical applications. But the essence of the Lin's constraints has been bewildered for more than half century.

Now considering a fluid particle, and applying Hamilton's principle to construct the following functional:

$$J(x_i) = \int_t \left(\frac{1}{2}\dot{x}_i^2 - \Pi\right) dt, \tag{5.269}$$

where dot means the partial derivation to time alone pathline, that is, material derivation, its Eulerian equation can be easily deduced

$$\ddot{x}_i = -\frac{\partial \Pi}{\partial x_i} = -\frac{1}{\rho}\frac{\partial P}{\partial x_i} \tag{5.270}$$

that is, the momentum equation for a perfect flow, and it is easy to prove [46] that $\Pi = E + P/\rho$, so Π represents enthalpy.

The involutory transformation is to convert Lagrange space into Euler space:

$$\dot{x}_i = u_i \quad (i = 1, 2, 3), \tag{5.271}$$

which are the pathline equations in the Eulerian form and are equivalent to (5.266) in Lagrangian form, that is, Lin's constraint equations. So it is very clear that Lin's constraints are actually the variational constraints introduced by the space exchange. The Hamilton principle is valid for Lagrange space, while all equations in fluid mechanics are derived through Euler space, and Lin's constraints arise in the space change.

Using the Lagrange multipliers to remove the constraints, (5.271), we have

$$J(x_i, u_i, \lambda_i) = \int_t \left(\frac{1}{2}v^2 - \Pi + \lambda_i(\dot{x}_i - u_i)\right) dt. \tag{5.272}$$

Identifying the Lagrange multipliers

$$\lambda_i = u_i \tag{5.273}$$

results in the following variational principle:

$$J(x_i, u_i) = \int_t \left(\frac{1}{2}v^2 - u_i\dot{x}_i + \Pi\right)dt. \tag{5.274}$$

Applying Hamilton's principle to a closed system, we have

$$J(x_i) = \int_t \int_{V(t)} \rho_0 \left(\frac{1}{2}\dot{x}_i^2 - E\right)dV(t)dt, \tag{5.275}$$

where $V(t)$ is the volume of the closed system in Lagrange space and keeps unchanged, and $\int_{V(t)} \rho_0 dV$ is the total mass in the closed system.

Making the involutory transformation, we obtain the following functional via Lagrange multiplier method

$$J(x_i, u_i) = \int_t \int_{V(t)} \rho_0 \left(u_i\dot{x}_i - \frac{1}{2}u_i^2 - E\right)dV(t)dt. \tag{5.276}$$

It is easy to see that the functional (5.275) or (5.276) ceases to be a stationary variational principle if we transform the closed system in Lagrange space to an open system is Euler space $\int_t \int_{V(t)} \rho_0 dV(t)dt \to \int_t \int_{CV} \rho dV dt$ where a new variable ρ in introduced.

In order to guarantee functional (5.276) to be still a stationary principle during space transformation, according to the semi-inverse method [14], let ρ be an independent variable and include an unknown function F in the functional:

$$J(u_i, \rho) = \int_t \int_{CV} \left\{\rho\left(\frac{1}{2}u_i^2 - E\right) + F(u_i, \rho)\right\}dV dt, \tag{5.277}$$

where F is an unknown functional of x_i, ρ, and CV is the control volume. In the next section the identification of F will be given in details.

Rotational equation with Clebsch variables can be expressed as follows:

$$\vec{v} = \nabla\Phi + \alpha\nabla\beta. \tag{5.278}$$

Assuming that $\partial\alpha/\partial t = 0$ and $D\alpha/Dt = 0$, we can deduce the following energy equation:

$$\frac{\partial\Phi}{\partial t} + \alpha\frac{\partial\beta}{\partial t} + \frac{1}{2}v^2 + \frac{\kappa}{\kappa - 1}\frac{P}{\rho} = R(t). \tag{5.279}$$

According to the semi-inverse method [17], we construct a trial functional in the form

$$J(\vec{v},\rho,\Pi,\Phi,\alpha,\beta) = \int_t\int_{CV}\left\{\rho\left(\frac{1}{2}v^2 - \Pi\right) + F\right\}dV dt, \tag{5.280}$$

where F is an unknown function of $\vec{v},\rho,\Pi,\Phi,\alpha,$ and β.

Making the functional (5.280) stationary with respect to $\vec{v}$, we have

$$\rho\vec{v} + \frac{\partial F}{\partial \vec{v}} = 0. \tag{5.281}$$

Assuming that (5.281) satisfies (5.278), we set

$$\frac{\partial F}{\partial \vec{v}} = -\rho\vec{v} = -\rho(\nabla\Phi + \alpha\nabla\beta). \tag{5.282}$$

The unknown function can be identified as follows:

$$F = -\rho\vec{v}\cdot(\nabla\Phi + \alpha\nabla\beta) + f, \tag{5.283}$$

where f is an unknown function of $\rho,\Pi,\Phi,\alpha,$ and β. Such and trial functional can be updated as follows:

$$J(\vec{v},\rho,\Pi,\Phi,\alpha,\beta) = \int_t\int_{CV}\left\{\rho\left(\frac{1}{2}v^2 - \vec{v}\cdot(\nabla\Phi + \alpha\nabla\beta) - \Pi\right) + f\right\}dV dt. \tag{5.284}$$

Now the stationary condition with respect to $\delta\Phi$ and continuity equation yield

$$\frac{\partial f}{\partial \Phi} = -\nabla\cdot(\rho\vec{v}) = \frac{\partial \rho}{\partial t}. \tag{5.285}$$

From (5.285), f can be determined in the form

$$f = \Phi\frac{\partial \rho}{\partial t} + h, \tag{5.286}$$

where h is an unknown function of $\rho,\Pi,\alpha,$ and β. Substituting (5.286) to the trial functional and then taking by part to yield the following new trial functional:

$$J(\vec{v},\rho,\Pi,\Phi,\alpha,\beta) = \int_t\int_{CV}\left\{\rho\left(\frac{1}{2}v^2 - \vec{v}\cdot(\nabla\Phi + \alpha\nabla\beta) - \Pi - \frac{\partial \Phi}{\partial t}\right) + h\right\}dV dt. \tag{5.287}$$

Doing the same as before,

$$\delta\rho : \frac{\partial h}{\partial \rho} = -\frac{1}{2}v^2 + \vec{v}\cdot(\nabla\Phi + \alpha\nabla\beta) + \Pi + \frac{\partial \Phi}{\partial t} = \frac{1}{2}v^2 + \Pi + \frac{\partial \Phi}{\partial t} = R - \alpha\frac{\partial \beta}{\partial t}. \tag{5.288}$$

We identifying the unknown function h to yield

$$h = \rho\left(R - \alpha\frac{\partial\beta}{\partial t}\right) + g, \tag{5.289}$$

where g is an unknown function of Π, α, and β, doing the same way as before,

$$\delta\Pi : \frac{\partial g}{\partial\Pi} = \rho = \left(\exp\left\{-\frac{S - S_0}{m}\right\}\frac{\kappa - 1}{\kappa}\Pi\right)^m. \tag{5.290}$$

Identifying the unknown function g to get

$$g = \left(\frac{\kappa - 1}{\kappa}\exp\left(-\frac{S - S_0}{m}\right)\right)^m \frac{\kappa - 1}{\kappa}\Pi^{\kappa/(\kappa-1)}, \tag{5.291}$$

we get the following generalized variational principle:

$$J(\vec{v}, \rho, \Pi, \Phi, \alpha, \beta) = \int_t\int_{CV} L\,dV\,dt, \tag{5.292}$$

where

$$\begin{aligned} L = {} & \rho\left(\frac{1}{2}v^2 - \vec{v}\cdot(\nabla\Phi + \alpha\nabla\beta) - \Pi - \frac{\partial\Phi}{\partial t} - \alpha\frac{\partial\beta}{\partial t} + R\right) \\ & + \left(\frac{\kappa - 1}{\kappa}\exp\left(-\frac{S - S_0}{m}\right)\right)^m \frac{\kappa - 1}{\kappa}\Pi^{\kappa/(\kappa-1)}. \end{aligned} \tag{5.293}$$

The deduced generalized variational principle (GVP) with only 6 independent variables, (5.292), is unknown to the present time and more general and concise than any other known generalized variationals such as the correspondent GVP with Lin's constraints which needs 11 independent variables ($\vec{v}, P, \rho$ and extra three Lin's constraints together with three Lagrange multipliers, and one Lagrange multiplier for continuity equation, one Lagrange multiplier for isoentropy equation) and actually is of little utility solving the 3D unsteady rotational flow, contrarily the obtained GVP of (5.292) is so beautiful and concise that the theorem makes it possible to deal with the problem via VP-based FEM; and have wide applicability of the solution ($DS/Dt \neq 0$). Making the above functional (5.292) stationary, we can get all the Eulerian equations and following additional two equations:

$$\begin{aligned} \frac{D\alpha}{Dt} &= 0, \\ \frac{D\beta}{Dt} &= \frac{1}{\rho}\left(\frac{\kappa - 1}{\kappa}\right)^m \frac{1}{\kappa}\Pi^{\kappa/(\kappa-1)}\exp\{-(S - S_0)\}\frac{\partial S}{\partial\alpha}, \\ &= \frac{1}{\kappa - 1}\rho^{\kappa-1}\exp\left\{-\frac{S - S_0}{m}\right\}\frac{\partial S}{\partial\alpha} = \frac{1}{\kappa}\frac{P}{\rho}\frac{\partial S}{\partial\alpha}. \end{aligned} \tag{5.294}$$

The first equation of (5.294) is actually Lin's constraint equation. So via semi-inverse method, the Lin's constraints equations can reduce to only one, which is the best way to deal with Lin's constraints in the history.

Hereby we have successfully explained the phenomenon of Lin's constraints via involutory transformation and deduced a generalized variational principle with only 6 independent variables via semi-inverse method, which makes it possible to use FEM to calculate the 3D unsteady compressible rotational flow. The corresponding generalized variational principle with Lin's constraints is actually an approximate one (for Herrivel's principle actually is a wrong principle), which, however, can find some application, especially in 2D problems, due to the fact that we can deduce all the Eulerian equations from the functional when making it stationary.

5.6. *Variational Iteration Method*

The variational iteration method [58–61] is an effective method for searching for various wave solutions including periodic solutions, solitons, and compacton solutions without linearization or weak nonlinearity assumptions, see, for example, [62–64].

5.6.1. The Lagrange Multiplier

The variational iteration method has been shown to solve a large class of nonlinear problems effectively, easily, and accurately with the approximations converging rapidly to accurate solutions.

To illustrate the basic idea of the technique, we consider the following general nonlinear system:

$$L[u(t)] + N[u(t)] = 0, \tag{5.295}$$

where L is a linear operator, and N is a nonlinear operator.

The basic concept of the method is to construct a correction functional for the system (5.295), which reads

$$u_{n+1}(t) = u_n(t) + \int_{t_0}^{t} \lambda\{Lu_n(s) + N\tilde{u}_n(s)\}ds, \tag{5.296}$$

where λ is a general Lagrange multiplier that can be identified optimally via variational theory, u_n is the nth approximate solution, and $\tilde{u}_n$ denotes a restricted variation, that is, $\delta\tilde{u}_n = 0$. To illustrate how restricted variation works in the variational iteration method, we consider a simple algebraic equation

$$x^2 - 3x + 2 = 0. \tag{5.297}$$

We rewrite (5.297) in the form

$$\tilde{x} \cdot x - 3x + 2 = 0, \tag{5.298}$$

where $\tilde{x}$ is called a restricted variable whose value is assumed to be known (5.6.1. the initial guess). Solving x from (5.298) leads to the result

$$x = \frac{2}{3 - \tilde{x}}, \tag{5.299}$$

or in iteration form,

$$x_{n+1} = \frac{2}{3 - x_n}. \tag{5.300}$$

After identifying the multiplier in (5.296), we have the following iteration algorithm:

$$u_{n+1}(t) = u_n(t) + \int_{t_0}^{t} \lambda\{Lu_n(s) + Nu_n(s)\}ds. \tag{5.301}$$

Equation (5.301) is called the variational iteration algorithm-I.

Consider the following nonlinear equation of kth order:

$$u^{(k)} + f\left(u, u', u'', \ldots, u^{(k)}\right) = 0. \tag{5.302}$$

The variational iteration formulation is constructed as follows:

$$u_{n+1}(t) = u_n(t) + \int_{t_0}^{t} \lambda\left(u_n^{(k)} + \tilde{f}_n\right)ds, \tag{5.303}$$

where $\delta\tilde{f}_n = 0$, $f_n = f(u_n, u_n', u_n'', \ldots)$. After identifying the multiplier, we have

$$u_{n+1}(t) = u_n(t) + (-1)^n \int_{t_0}^{t} \frac{1}{(n-1)!}(s-t)^{n-1}\left[u_n^{(k)}(s) + f_n\right]ds. \tag{5.304}$$

This formulation is generally called the variational iteration algorithm-I [11, 61].

The main merit of this iteration formula lies in the fact that $u_0(t)$, the initial solution, can be freely chosen, with even unknown parameters contained. However, some repeated and unnecessary iterations are involved in this iteration algorithm at each step.

For initial value problems, we can begin with

$$u_0(t) = u(0) + tu'(0) + \frac{1}{2!}t^2u''(0) + \cdots \frac{1}{k!}t^k u^{(k)}(0). \tag{5.305}$$

This leads to a series solution converging to the exact solution.

For boundary value problems, the initial guess can be expressed in the form

$$u_0(t) = a_1g_1(t) + a_2g_2(t) + \cdots + a_kg_k(t), \tag{5.306}$$

where $g_k(t)$ are known functions, and a_k are unknowns to be further determined after a few iterations by the boundary conditions.

After identifying the Lagrange multiplier λ in (5.296), we can construct the iteration formula

$$u_{n+1}(t) = u_0(t) + (-1)^n \int_{t_0}^{t} \frac{1}{(n-1)!}(s-t)^{n-1} f_n ds, \tag{5.307}$$

instead of the iteration algorithm (5.301). We call (5.307) the variational iteration algorithm-II [11, 61].

Note: u_0 must satisfy the initial/boundary conditions. This is the main shortcoming of the algorithm.

From (5.307), we obtain the following variational iteration algorithm-III

$$u_{n+2}(t) = u_{n+1}(t) + \int_{t_0}^{t} \lambda\{f_{n+1}(s) - f_n(s)\} ds. \tag{5.308}$$

One common property of both the variational iteration algorithm-I and the variational iteration algorithm-III is the allowed dependence of the initial guess on unknown parameters whose values could be identified after a few iterations by using the initial/boundary conditions. the variational iteration algorithm-III, in particular, is highly suitable for boundary value problems of high orders.

5.6.2. Laplace Transform for Identification of the Lagrange Multiplier [65]

We can also apply Laplace transform to identify the Lagrange multiplier. By Laplace transform, we have

$$s^k U(s) + \ell\left\{f\left(u, u', \ldots, u^{(k)}\right)\right\} = 0, \tag{5.309}$$

or

$$U(s) = \frac{-\ell\{f(u, u', \ldots, u^{(k)})\}}{s^k}. \tag{5.310}$$

The inverse Laplace transform reads

$$u(t) = (-1)^k \int_0^t \frac{(\eta - t)^{k-1}}{(k-1)!} f\left(u_n(\eta), u_n'(\eta), \ldots, u_n^{(k)}(\eta)\right) d\eta. \tag{5.311}$$

Hence, the following iteration algorithm is derived:

$$u_{n+1}(t) = u_0(t) + (-1)^k \int_0^t \frac{(\eta - t)^{k-1}}{(k-1)!} f\left(u_n(\eta), u_n'(\eta), \ldots, u_n^{(k)}(\eta)\right) d\eta. \tag{5.312}$$

5.6.3. Variational Iteration Method for Solitary Solutions

Here is an incomplete list for variational iteration formulas for various differential equations [8, 9]:

$$
\begin{gathered}
u' + f(u,u') = 0, \\
u_{n+1}(t) = u_0(t) - \int_0^t f(u_n, u_n')ds, \\
u' + \alpha u + f(u,u') = 0, \\
u_{n+1}(t) = u_0(t) - \int_0^t e^{\alpha(s-t)} f(u_n, u_n')ds, \\
u'' + f(u,u',u'') = 0, \\
u_{n+1}(t) = u_0(t) + \int_0^t (s-t) f(u_n, u_n', u_n'')ds, \\
u'' + \omega^2 u + f(u,u',u'') = 0, \\
u_{n+1}(t) = u_0(t) + \frac{1}{\omega}\int_0^t \sin\omega(s-t) f(u_n, u_n', u_n'')ds, \\
u'' - \alpha^2 u + f(u,u',u'') = 0, \\
u_{n+1}(t) = u_0(t) + \int_0^t \frac{1}{2\alpha}\left(e^{\alpha(s-t)} - e^{\alpha(t-s)}\right) f(u_n, u_n', u_n'')ds, \\
u''' + f(u,u',u'',u''') = 0, \\
u_{n+1}(t) = u_0(t) - \int_0^t \frac{1}{2}(s-t)^2 f(u_n, u_n', u_n'', u_n''')ds, \\
u^{(4)} + f\left(u,u',u'',u''',u^{(4)}\right) = 0, \\
u_{n+1}(t) = u_0(t) + \int_0^t \frac{1}{6}(s-t)^3 f\left(u_n, u_n', u_n'', u_n''', u_n^{(4)}\right)ds, \\
u^{(k)} + f\left(u,u',u'',\dots,u^{(k)}\right) = 0, \\
u_{n+1}(t) = u_0(t) + (-1)^n \int_0^t \frac{1}{(n-1)!}(s-t)^{n-1} f\left(u_n, u_n', u_n'', \dots, u_n^{(k)}\right)ds.
\end{gathered}
\tag{5.313}
$$

The main feature of the method is that the initial solution can be chosen with some unknown parameters in the form of the searched solution. For example, for solitons, we begin with

$$u_0 = p\,\mathrm{sech}^2(q\xi), \quad \xi = x + ct, \tag{5.314}$$

where p and q are the unknown parameters to be further identified after one or few iterations. For a more general form for solitary solutions, we assume that

$$u_0(\xi) = \frac{\sum_{i=-n}^{m} c_i e^{i\xi}}{\sum_{i=-p}^{q} b_i e^{i\xi}}, \quad \xi = x + ct, \tag{5.315}$$

where c_i and b_i are constants to be further determined.

For discontinuous solitons, we can assume, for example, the following form:

$$u(\xi) = p\exp(-q|\xi|), \tag{5.316}$$

or

$$u_0(\xi) = \frac{\sum_{i=-n}^{m} c_i e^{i|\xi|}}{\sum_{i=-p}^{q} b_i e^{i|\xi|}}, \tag{5.317}$$

where p and q are the unknown parameters to be further identified.

For compacton-like solution, we assume that the solution has the form

$$u_0(x,t) = \frac{a\sin^2(kx+\omega t)}{b + c\sin^2(kx+\omega t)}, \tag{5.318}$$

where $a, b, k,$ and ω are unknown constants further to be determined

$$u_0(x,t) = \frac{\sum_i a_i\sin^2(k_i x+\omega_i t)}{\sum_j \left[b_j + c_j\sin^2(k_j x+\omega_j t)\right]}. \tag{5.319}$$

As an illustrating example, we consider the following modified KdV equation

$$u_t + u^2 u_x + u_{xxx} = 0. \tag{5.320}$$

Its iteration formulation can be constructed as follows:

$$u_{n+1}(x,t) = u_n(x,t) - \int_0^t \left\{(u_n)_t + u_n^2 (u_n)_x + (u_n)_{xxx}\right\} dt. \tag{5.321}$$

To search for its compacton-like solution, we assume that the solution has the form

$$u_0(x,t) = \frac{a\sin^2(kx+wt)}{b + c\sin^2(kx+wt)}, \tag{5.322}$$

where a, b, k,and w are unknown constants further to be determined.

Substituting (5.322) into (5.321), we can calculate u_1 and u_2 with ease. In order to identify the constants in the initial solution, we can set

$$\frac{\partial^k}{\partial t^k} u_n(x,t) = \frac{\partial^k}{\partial t^k} u_{n+1}(x,t), \tag{5.323}$$

setting

$$\frac{\partial}{\partial t} u_0(x,t) = \frac{\partial}{\partial t} u_1(x,t). \tag{5.324}$$

By a simple calculation (see [60] for details), the compacton-like solution is obtained, which reads

$$u(x,t) = \frac{\pm 2\sqrt{2}kc \sin^2(kx - 4k^3 t)}{-(3/2)c + c\sin^2(kx - 4k^3 t)}, \tag{5.325}$$

or

$$u(x,t) = \frac{\pm 4\sqrt{2}k \sin^2(kx - 4k^3 t)}{-3 + 2\sin^2(kx - 4k^3 t)}. \tag{5.326}$$

We can also begin with a more general initial solution in the form

$$u_0(x,t) = a + \frac{1}{c + d\cos(kx + wt)}, \tag{5.327}$$

where a, b, c, k, and w are unknown constants.

Setting $\partial u_0/\partial t = \partial u_1/\partial t$, we, therefore, obtain the following new compacton-like solution:

$$u(x,t) = a + \frac{1}{-a/3k^2 \pm (1/6k^2)\sqrt{4a^2 - 6k^2}\cos(kx + (k^3 - ka^2)t)}. \tag{5.328}$$

If we choose $k = iK$, where K is a constant, then (5.328) becomes

$$\begin{aligned} u(x,t) &\\ &= a + \frac{1}{(a/3K^2) \pm (1/6K^2)\sqrt{4a^2 + 6K^2}\cosh(Kx + (-K^3 - Ka^2)t)} \\ &= a + \frac{1}{(a/3K^2) \pm (1/12K^2)\sqrt{4a^2 + 6K^2}\left[\exp(Kx + (-K^3 - Ka^2)t) + \exp(-Kx - (-K^3 - Ka^2)t)\right]}, \end{aligned} \tag{5.329}$$

which is a solitary solution. It is interesting that we can convert the compacton-like solution to solitary solution.

The initial solution (trial function) can be also constructed in a solitary form. Now we begin with

$$u_0 = a + \frac{1}{b + c\exp(kx + wt) + d\exp(-kx - wt)}. \tag{5.330}$$

By the same manipulation as illustrated above, we set $\partial u_0/\partial t = \partial u_1/\partial t$.

So we obtain the following needed solitary solution:

$$u(x,t) = a + \frac{1}{a/3k^2 + c\exp(kx + (-k^3 - ka^2)t) + ((3k^2 + 2a^2)/72k^4)(1/c)\exp(-kx - (-k^3 - ka^2)t)}. \tag{5.331}$$

Hereby a and c are free parameters.

It is also interesting to note that the solitary solution can be converted into a compacton-like solution if we choose $k = iK$, where K is a constant. If $k = iK$, (5.331) becomes

$$u(x,t) = a + \frac{1}{-a/3K^2 + (c + \mathcal{A})\cos(Kx + (K^3 - Ka^2)t) + (c - \mathcal{A})i\sin(Kx + (K^3 - Ka^2)t)}, \tag{5.332}$$

where $\mathcal{A}$ donates $(-3K^2 + 2a^2)/72K^4c$. In the above derivation, we use the relations

$$\begin{aligned} \exp(ikx + iwt) &= \cos(kx + wt) + i\sin(kx + wt), \\ \exp(-ikx - iwt) &= \cos(kx + wt) - i\sin(kx + wt). \end{aligned} \tag{5.333}$$

In order to convert (5.332) into a compact form, the last term in denominator must be vanished, which requires

$$c - \frac{-3K^2 + 2a^2}{72K^4c} = 0. \tag{5.334}$$

Solving c from (5.334) results in

$$c = \pm\frac{1}{12}\sqrt{\frac{-6K^2 + 4a^2}{K^2}}. \tag{5.335}$$

Substituting (5.335) into (5.332) yields a compacton-like solution

$$u(x,t) = a + \frac{1}{-a/3K^2 \pm (1/6K^2)\sqrt{4a^2 - 6K^2}\cos(Kx + (K^3 - ka^2)t)}. \tag{5.336}$$

5.7. Homotopy Perturbation Method

5.7.1. Homotopy Technology

Before proceeding the method, we give an interesting application of the homotopy technology to an asymptotic match of Rayleigh grain size distribution and Hillert grain size distribution. Grain growth is a well-known phenomenon in the evolution of crystalline microstructures. A stochastic continuity or Fokker-Planck continuity equation was proposed to accurately predict properties of grain growth [66]. The equation can be expressed in the form [66]

$$\varepsilon\frac{d^2 f}{dx^2} + \left(\frac{\alpha+\varepsilon}{x} - \alpha + 2x\right)\frac{df}{dx} + \left(6 - \frac{\alpha+\varepsilon}{x^2}\right)f = 0. \tag{5.337}$$

Let us consider two limiting cases where the driving force is either due to diffusion (Rayleigh) or to the drift velocity (Hillert). In the first case, the Rayleigh grain size distribution occurs when $\alpha = 0$ and $\varepsilon \neq 0$ [66]:

$$f_r(x) = \frac{\pi}{2}x\exp\left(-\frac{\pi}{4}x^2\right). \tag{5.338}$$

The other limiting case, $\alpha = 8$ and $\varepsilon = 0$, leads to Hillert grain size distribution [66]:

$$f_h(x) = \frac{8\exp(2x/(x-2))}{(x-2)^2}. \tag{5.339}$$

The actual solution to (5.337) falls between (5.338) and (5.339); in the work of Pande and Cooper [66], a very good but approximate solution was given (see (18) of Pande and Cooper's publication [66]). Hereby we will suggest a simple matching technology to bridge the limiting cases.

An asymptotic match of (5.338) and (5.339) can be expressed in the form [67]

$$\begin{aligned} f(x,\alpha,\varepsilon) &= C(\alpha,\varepsilon)f_r(x) + D(\alpha,\varepsilon)f_h(x) \\ &= C(\alpha,\varepsilon)\frac{\pi}{2}x\exp\left(-\frac{\pi}{4}x^2\right) + D(\alpha,\varepsilon)\frac{8\exp(2x/(x-2))}{(x-2)^2}, \end{aligned} \tag{5.340}$$

where C and D are matching parameters.

Equation (5.340) should turn out to be exactly (5.338) and (5.339), respectively, for the above two limiting cases. This requires

$$\begin{aligned} C(0,\varepsilon) &= 1, \qquad D(0,\varepsilon) = 0, \\ C(8,0) &= 0, \qquad D(8,0) = 1. \end{aligned} \tag{5.341}$$

We can freely choose C and D satisfying (5.341) and the following normalization condition

$$C + D = 1. \tag{5.342}$$

Hereby we give simple expressions for C and D in the forms

$$C(\alpha,\varepsilon) = 1 - \frac{1}{8}\alpha(1-\varepsilon),$$
$$D(\alpha,\varepsilon) = \frac{1}{8}\alpha(1-\varepsilon). \tag{5.343}$$

We, therefore, obtain the following approximate solution to (5.337) [67]:

$$f(x) = \frac{\pi}{2}\left(1 - \frac{1}{8}\alpha(1-\varepsilon)\right)x\exp\left(-\frac{\pi}{4}x^2\right) + \frac{1}{8}\alpha(1-\varepsilon)\frac{8\exp(2x/(x-2))}{(x-2)^2}. \tag{5.344}$$

This equation is simpler than that given by Pande and Cooper [66]. There are many alternative ways for determination of C and D satisfying (5.341) and (5.342).

In this short paper we suggest an asymptotic match to bridge the two limited cases, and the obtained result is valid for the whole case. The asymptotic match is a simple and useful mathematical tool in engineering for reliable treatment of a nonlinear problem whose analytical solution can be easily obtained for two limited cases (e.g., $\varepsilon \to 0$ and $\varepsilon \to \infty$, resp.) by some analytical methods.

Consider another example of the relativistic oscillator

$$u'' + \frac{u}{\sqrt{1+u^2}} = 0, \tag{5.345}$$

with initial conditions $u(0) = A$, $u'(0) = 0$.

It is easy to obtain the following approximate frequency [68]:

$$\omega = \frac{2\sqrt{2}}{\pi}\left(1 + \frac{4}{\pi^2}\cdot A^2\right)^{-1/4}, \quad \text{for } A \gg 1,$$
$$\omega = \left(1 + \frac{1}{2}A^2\right)^{-1/4}, \quad \text{for } A \ll 1. \tag{5.346}$$

In order to match both the cases $A \to 0$ and $A \to \infty$, we construct the following homotopy [68]:

$$\tilde{\omega} = e^{-\alpha A}f(A) + \left(1 - e^{-\alpha A}\right)g(A), \tag{5.347}$$

where $f(A) = (2\sqrt{2}/\pi)(1 + (4/\pi^2)\cdot A^2)^{-1/4}$, $g(A) = (1 + (1/2)A^2)^{-1/4}$, and α is a free parameter. Now considering the case when $A = 1$, we have exact frequency, which is $\tilde{\omega} = 0.8736$. From this relationship, we can identify α as follows:

$$\alpha = 0.4962. \tag{5.348}$$

Finally, we obtain the following result:

$$\tilde{\omega} = e^{-0.4962A} \cdot \frac{2\sqrt{2}}{\pi}\left(1 + \frac{4}{\pi^2} \cdot A^2\right)^{-1/4} + \left(1 - e^{-0.496A}\right) \cdot \left(1 + \frac{1}{2}A^2\right)^{-1/4}. \tag{5.349}$$

5.7.2. Homotopy Perturbation Method with an Auxiliary Term [69]

The two most important steps in application of the homotopy perturbation method [69–72] are to construct a suitable homotopy equation and to choose a suitable initial guess. The homotopy equation should be constructed such that, when the homotopy parameter is zero, it can approximately describe the solution property, and the initial solution can be chosen with an unknown parameter, which is determined after one or two iterations. This paper suggests an alternative approach for construction of the homotopy equation with an auxiliary term Duffing equation is used as example to illustrate the solution procedure.

Consider a general nonlinear equation

$$Lu + Nu = 0, \tag{5.350}$$

where L and N are, respectively, the linear operator and nonlinear operator.

The first step for the method is to construct a homotopy equation in the form

$$\tilde{L}u + p\left(Lu - \tilde{L}u + Nu\right) = 0, \tag{5.351}$$

where $\tilde{L}$ is a linear operator with a possible unknown constant, and $\tilde{L}u = 0$ can best describe the solution property. The embedding parameter p monotonically increases from zero to unit as the trivial problem, $\tilde{L}u = 0$, is continuously deformed to the original one.

For example, consider a nonlinear oscillator [9]

$$u'' + \varepsilon u^3 = 0, \qquad u(0) = A, \qquad u'(0) = 0. \tag{5.352}$$

For an oscillator, we can use sine or cosine function. We assume that the approximate solution of (5.352) is

$$u(t) = A\cos\omega t, \tag{5.353}$$

where ω is the frequency to be determined later. We, therefore, can choose

$$\tilde{L}u = \ddot{u} + \omega^2 u. \tag{5.354}$$

Accordingly, we can construct a homotopy equation in the form

$$\ddot{u} + \omega^2 u + p\left(u^3 - \omega^2 u\right) = 0. \tag{5.355}$$

When $p = 0$, we have

$$\ddot{u} + \omega^2 u = 0, \qquad u(0) = A, \qquad u'(0) = 0, \tag{5.356}$$

which describes the basic solution property of the original nonlinear equation (5.352).

When $p = 1$, (5.355) becomes the original one. So the solution procedure is to deform from the initial solution, (5.353), to the real one. Due to one unknown parameter in the initial solution, only one iteration is enough. For detailed solution procedure, refer to [9].

If a higher-order approximate solution is searched for, we can construct a homotopy equation in the form

$$\ddot{u} + 0 \cdot u + pu^3 = 0. \tag{5.357}$$

We expand the solution and the coefficient, zero, of the linear term into a series in p:

$$u = u_0 + pu_1 + p^2 u_2 + \cdots, \tag{5.358}$$

$$0 = \omega^2 + pa_1 + p^2 a_2 + \cdots, \tag{5.359}$$

where the unknown constant, a_i, is determined in the $(i + 1)$th iteration. The solution procedure is given in [9].

In this paper, we suggest an alternative approach for construction of homotopy equation, which is

$$\tilde{L}u + p\left(Lu - \tilde{L}u + Nu\right) + \alpha p(1 - p)u = 0, \tag{5.360}$$

where α is an auxiliary parameter. When $\alpha = 0$, (5.360) turns out to be that of the classical one, expressed in (5.351). The auxiliary term, $\alpha p(1 - p)u$, vanishes completely when $p = 0$ or $p = 1$, so the auxiliary term will affect neither the initial solution ($p = 0$), nor the real solution ($p = 1$).

To illustrate the solution procedure, we consider a nonlinear oscillator in the form

$$\frac{d^2u}{dt^2} + bu + cu^3 = 0, \qquad u(0) = A, \qquad u'(0) = 0, \tag{5.361}$$

where b and c are positive constants.

Equation (5.361) admits a periodic solution, and the linearized equation of (5.361) is

$$u'' + \omega^2 u = 0, \qquad u(0) = A, \qquad u'(0) = 0, \tag{5.362}$$

where ω is the frequency of (5.361).

We construct the following homotopy equation with an auxiliary term [69]:

$$u'' + \omega^2 u + p\left[\left(b - \omega^2\right)u + cu^3\right] + \alpha p(1 - p)u = 0. \tag{5.363}$$

Assume that the solution can be expressed in a power series in p as shown in (5.358). Substituting (5.358) into (5.363), and processing as the standard perturbation method, we have

$$u_0'' + \omega^2 u_0 = 0, \qquad u_0(0) = A, \qquad u_0'(0) = 0, \tag{5.364}$$

$$u_1'' + \omega^2 u_1 + \left(b - \omega^2\right) u_0 + c u_0^3 + \alpha u_0 = 0, \qquad u_1(0) = 0, \qquad u_1'(0) = 0, \tag{5.365}$$

$$u_2'' + \omega^2 u_2 + \left(b - \omega^2\right) u_1 + 3 c u_0^2 u_1 + \alpha(u_1 - u_0) = 0, \tag{5.366}$$

with initial conditions

$$\sum_{i=0} u_i(0) = A, \qquad \sum_{i=0} u_i'(0) = 0. \tag{5.367}$$

Solving (5.364), we have

$$u_0 = A \cos \omega t. \tag{5.368}$$

Substituting u_0 into (5.365) results in

$$u_1'' + \omega^2 u_1 + A\left(\alpha + b - \omega^2 + \left(\frac{3}{4}\right) c A^2\right) \cos \omega t + \frac{1}{4} c A^3 \cos 3\omega t = 0. \tag{5.369}$$

Eliminating the secular term needs

$$\alpha + b - \omega^2 + \frac{3}{4} c A^2 = 0. \tag{5.370}$$

A special solution of (5.369) is

$$u_1 = -\frac{c A^3}{32 \omega^2} \cos 3\omega t. \tag{5.371}$$

If only a first-order approximate solution is enough, we just set $\alpha = 0$, and this results in

$$\omega = \sqrt{b + \frac{3}{4} c A^2}. \tag{5.372}$$

The accuracy reaches 7.6% even for the case $cA^2 \to \infty$.

The solution procedure continues by submitting u_1 into (5.366), and after some simple calculation, we obtain

$$u_2'' + \omega^2 u_2 - \left(\alpha A + \frac{3c^2A^5}{128\omega^2}\right)\cos\omega t - \left(\frac{cA^3(b-\omega^2)}{32\omega^2} + \frac{3c^2A^5}{64\omega^2} + \frac{\alpha cA^3}{32\omega^2}\right)\cos 3\omega t - \frac{3c^2A^5}{128\omega^2}\cos 5\omega = 0. \tag{5.373}$$

No secular term in u_2 requires

$$\alpha A + \frac{3c^2A^5}{128\omega^2} = 0. \tag{5.374}$$

Solving (5.370) and (5.374) simultaneously, we obtain

$$\omega = \sqrt{\frac{b + (3/4)cA^2 + \sqrt{(b+(3/4)cA^2)^2 + (3/32)c^2A^4}}{2}}, \tag{5.375}$$

and the approximate solution is $u(t) = A\cos\omega t$, where ω is given in (5.375).

In order to compare with the perturbation solution and the exact solution, we set $b = 1$. In case $c \ll 1$, (5.375) agrees with that obtained by the classical perturbation method; when $c \to \infty$, we have

$$\lim_{c\to\infty}\omega = \sqrt{\frac{3/4 + \sqrt{(3/4)^2 + 3/32}}{2}}\sqrt{cA^2} = 0.8832\sqrt{cA^2}. \tag{5.376}$$

The exact period reads

$$T_{ex} = \frac{4}{\sqrt{1+cA^2}}\int_0^{\pi/2}\frac{dx}{\sqrt{1-k\sin^2 x}}, \tag{5.377}$$

where $k = cA^2/2(1+cA^2)$.

In case $c \to \infty$, we have

$$\lim_{c\to\infty} T_{ex} = \frac{6.743}{\sqrt{cA^2}}, \tag{5.378}$$

$$\omega_{ex} \approx \frac{2\pi}{6.743}\sqrt{cA^2}. \tag{5.379}$$

Comparing between (5.376) and (5.378), we find that the accuracy reaches 5.5%, while accuracy of the first-order approximate frequency is 7.6%.

If a higher-order approximate solution is needed, we rewrite the homotopy equation in the form

$$u'' + \omega^2 u + p\left[\left(b - \omega^2\right)u + cu^3\right] + 1 \cdot p(1 - p)u = 0. \tag{5.380}$$

The coefficient, 1, in the auxiliary term is also expanded in a series in p in the form

$$1 = \alpha_0 + p\alpha_1 + p^2\alpha_2 + \cdots, \tag{5.381}$$

where α_i is identified in the $(i + 2)$th iteration. The solution procedure is similar to that illustrated above.

Generally, the homotopy equation can be constructed in the form

$$\tilde{L}u + p\left(Lu - \tilde{L}u + Nu\right) + 1 \cdot f(p)g(p)h(u, u', u'', \dots) = 0, \tag{5.382}$$

where f and g are functions of p, satisfying $f(0) = 0$ and $g(1) = 0$, and h can be generally expressed in the form

$$h = u + \beta_1 u' + \beta_2 u'' + \cdots. \tag{5.383}$$

5.7.3. Homotopy Perturbation Method for Solitary Solutions

Homotopy perturbation method [69–72] provides a simple mathematical tool for searching for soliton solutions without any small perturbation. Considering the following nonlinear equation:

$$\frac{\partial u}{\partial t} + au\frac{\partial u}{\partial x} + b\frac{\partial^3 u}{\partial x^3} + N(u) = 0, \quad a > 0,\ b > 0, \tag{5.384}$$

we can construct a homotopy in the form

$$(1 - p)\left\{\frac{\partial u}{\partial t} + 6u\frac{\partial u}{\partial x} + \frac{\partial^3 u}{\partial x^3}\right\} + p\left\{\frac{\partial u}{\partial t} + au\frac{\partial u}{\partial x} + b\frac{\partial^3 u}{\partial x^3} + N(u)\right\} = 0. \tag{5.385}$$

When $p = 0$, we have

$$\frac{\partial u}{\partial t} + 6u\frac{\partial u}{\partial x} + \frac{\partial^3 u}{\partial x^3} = 0, \tag{5.386}$$

a well-known KdV equation, whose soliton solution is known. When $p = 1$, (5.385) turns out to be the original equation. According to the homotopy perturbation method, we assume that

$$u = u_0 + pu_1 + p^2 u_2 + \cdots. \tag{5.387}$$

Substituting (5.387) into (5.385), and proceeding the same process as the traditional perturbation method does, we can easily solve u_0, u_1, and other components. The solution can be expressed finally in the form

$$u = u_0 + u_1 + u_2 + \cdots . \tag{5.388}$$

The homotopy perturbation method always stops before the second iteration, so the solution can be expressed as

$$u = u_0 + u_1, \tag{5.389}$$

for most cases.

For traveling wave solutions, we can use the following transformation:

$$u(t, x) = u(\xi), \quad \xi = x - Dt. \tag{5.390}$$

Equation (5.384) becomes

$$-Du_\xi + auu_\xi + bu_{\xi\xi\xi} + N(u) = 0. \tag{5.391}$$

Integrating the equation with respect to ξ, we have

$$-Du + \frac{1}{2}au^2 + bu_{\xi\xi} + F(u) = 0, \tag{5.392}$$

where $\partial F/\partial u = N(u) + C$, and C is an integral constant.

We begin with a soliton in the form

$$u_0 = \frac{1}{e^{-\alpha\xi} + e^{\alpha\xi}}, \tag{5.393}$$

which is the solution of the following equation:

$$u_{\xi\xi} - \alpha^2 u + 8\alpha^2 u^3 = 0. \tag{5.394}$$

Accordingly, we construct the following homotopy equation:

$$u_{\xi\xi} - \frac{D}{b}u + 0 \cdot u^3 + p\left[\frac{a}{2b}u^2 + \frac{1}{b}F(u)\right] = 0. \tag{5.395}$$

The solution and the parameters D/b and 0 are expanded in the forms

$$\begin{aligned} u &= u_0 + pu_1 + p^2u_2 + \cdots, \\ \frac{D}{b} &= \alpha^2 + p\alpha_1 + p^2\alpha_2 + \cdots, \\ 0 &= 8\alpha^2 + p\beta_1 + p^2\beta_2 + \cdots. \end{aligned} \tag{5.396}$$

Substituting (5.396) into (5.395), and proceeding a similar way as the perturbation method, we have

$$\begin{aligned} &u_{0\xi\xi} - \alpha^2 u_0 + 8\alpha^2 u_0^3 = 0, \\ &u_{1\xi\xi} - \alpha^2 u_1 + \alpha_1 u_0 + 24\alpha^2 u_0^2 u_1 + \beta_1 u_0^3 + \frac{a}{2b}u_0^2 + \frac{1}{b}F(u_0) = 0. \end{aligned} \tag{5.397}$$

As an example, we consider the following equation:

$$u_t + uu_x + \eta u_{xxx} = 0. \tag{5.398}$$

By the transformation $u(t,x) = u(\xi)$, $\xi = x - Dt$, we have

$$-Du_\xi + uu_\xi + \eta u_{\xi\xi\xi} = 0, \tag{5.399}$$

or

$$u'' - \frac{D}{\eta}u + \frac{1}{2\eta}u^2 + C = 0. \tag{5.400}$$

In order to make the solution process simple, we construct a homotopy equation in the form

$$u'' - \frac{D}{\eta}u + p\left(\frac{1}{2\eta}u^2 + C\right) = 0. \tag{5.401}$$

Assume that the solution and the parameter D/η can be expanded in the forms

$$\begin{aligned} u &= u_0 + pu_1 + p^2u_2 + \cdots, \\ \frac{D}{\eta} &= \lambda_0^2 + p\lambda_1 + p^2\lambda_2 + \cdots. \end{aligned} \tag{5.402}$$

Substituting (5.402) into (5.401), proceeding as the classical perturbation method does, we have

$$\begin{aligned} &u_0'' - \lambda_0^2 u_0 = 0, \\ &u_1'' - \lambda_0^2 u_1 - \lambda_1 u_0 + \frac{1}{2\eta} u_0^2 + C = 0, \\ &u_2'' - \lambda_0^2 u_2 - \lambda_2 u_0 - \lambda_1 u_1 + \frac{1}{\eta} u_0 u_1 = 0. \end{aligned} \tag{5.403}$$

The solution of the first equation of (5.403) is

$$u_0 = Ae^{\lambda_0 \xi} + Be^{-\lambda_0 \xi}. \tag{5.404}$$

Substituting (5.404) into the second equation of (5.403) we have

$$\begin{aligned} u_1'' - \lambda_0^2 u_1 &= \lambda_1 \left(Ae^{\lambda_0 \xi} + Be^{-\lambda_0 \xi} \right) - \frac{1}{2\eta} \left(Ae^{\lambda_0 \xi} + Be^{-\lambda_0 \xi} \right)^2 - C \\ &= \lambda_1 \left(Ae^{\lambda_0 \xi} + Be^{-\lambda_0 \xi} \right) - \frac{1}{2\eta} \left(A^2 e^{2\lambda_0 \xi} + B^2 e^{-2\lambda_0 \xi} \right) - \frac{AB}{\eta} - C. \end{aligned} \tag{5.405}$$

We set

$$\lambda_1 = 0, \qquad AB = -\eta C \tag{5.406}$$

in (5.405) to avoid the secular-like terms $\xi^k e^{\lambda_0 \xi}$, $\xi^k e^{-\lambda_0 \xi}$ or ξ^k $(k \geq 0)$, and solving the resultant equation, we have the solution for u_1:

$$u_1 = -\frac{1}{6\eta} \left(A^2 e^{2\lambda_0 \xi} + B^2 e^{-2\lambda_0 \xi} \right). \tag{5.407}$$

Substituting u_0 and u_1 into the third equation of (5.403) results in

$$\begin{aligned} u_2'' - \lambda_0^2 u_2 &= \lambda_2 \left(Ae^{\lambda_0 \xi} + Be^{-\lambda_0 \xi} \right) + \frac{1}{6\eta^2} \left(Ae^{\lambda_0 \xi} + Be^{-\lambda_0 \xi} \right) \left(A^2 e^{2\lambda_0 \xi} + B^2 e^{-2\lambda_0 \xi} \right) \\ &= A \left(\lambda_2 + \frac{AB}{6\eta^2} \right) e^{\lambda_0 \xi} + B \left(\lambda_2 + \frac{AB}{6\eta^2} \right) e^{-\lambda_0 \xi} + \frac{1}{6\eta^2} \left(A^3 e^{3\lambda_0 \xi} + B^3 e^{-3\lambda_0 \xi} \right). \end{aligned} \tag{5.408}$$

Similarly, we set

$$\lambda_2 = -\frac{AB}{6\eta^2}. \tag{5.409}$$

The solution for u_2 is

$$u_2 = \frac{1}{48\eta^2}\left(A^3 e^{3\lambda_0\xi} + B^3 e^{-3\lambda_0\xi}\right). \tag{5.410}$$

If the second order approximate solution is enough, then we have

$$u = u_0 + u_1 + u_2 = Ae^{\lambda_0\xi} + Be^{-\lambda_0\xi} - \frac{1}{6\eta}\left(A^2 e^{2\lambda_0\xi} + B^2 e^{-2\lambda_0\xi}\right) + \frac{1}{48\eta^2}\left(A^3 e^{3\lambda_0\xi} + B^3 e^{-3\lambda_0\xi}\right). \tag{5.411}$$

Solving λ_0 from the second equation of (5.402), we have

$$\lambda_0 = \sqrt{\frac{D}{\eta} - \left(p\lambda_1 + p^2\lambda_2 + \cdots\right)}\Bigg|_{p=1,\lambda_i=0(i\geq 3)} = \sqrt{\frac{D}{\eta} + \frac{AB}{6\eta^2}}. \tag{5.412}$$

Using in the $[1,2]$ exponential Padé approximant [62], we have

$$u = \frac{Ae^{\lambda_0\xi} + Be^{-\lambda_0\xi}}{1 + ae^{\lambda_0\xi} + be^{-\lambda_0\xi} + ce^{2\lambda_0\xi} + de^{-2\lambda_0\xi}}, \tag{5.413}$$

where the parameters $A, B, a, b, c,$ and d can be identified by the exponential Padé approximant similar to the Padé approximant.

The $[m,n]$ exponential Padé approximant is defined as follows.

Let

$$F = F_0 + \varepsilon F_1 + \varepsilon^2 F_2 + \cdots, \quad \varepsilon \longrightarrow 0. \tag{5.414}$$

The $[m,n]$ exponential Padé approximant reads

$$F_{mn}(\varepsilon) = \frac{\sum_{i=0}^{m} c_i \exp d_i\varepsilon}{\sum_{i=0}^{n} a_i \exp b_i\varepsilon}. \tag{5.415}$$

The coefficients $a_i, b_i, c_i,$ and d_i, are determined from the following condition: the first $(2m+2n)$ components of the expansion of the rational function $F_{mn}(\varepsilon)$ in a Maclaurin series coincide with the first $(2m+2n)$ components of the series $F(\varepsilon)$.

5.7.4. Couple Homotopy Perturbation Method with Laplace Transform [73]

Couple of the homotopy perturbation method with the Laplace transform makes the solution procedure much simpler.

Consider the following foam drainage equation:

$$u_t + 2u^2 u_x - u_x^2 - \frac{1}{2}u_{xx}u = 0, \tag{5.416}$$

with initial conditions

$$u(x,0) = -\frac{1}{2} + \frac{1}{1+e^x}. \tag{5.417}$$

Apply Laplace transformation

$$u(x,s) = \frac{1}{s}\left(-\frac{1}{2} + \frac{1}{1+e^x}\right) - \frac{1}{s}L\left(2u^2u_x - u_x^2 - \frac{1}{2}u_{xx}u\right). \tag{5.418}$$

Apply inverse Laplace transformation

$$u(x,t) = \left(-\frac{1}{2} + \frac{1}{1+e^x}\right) - L^{-1}\left(\frac{1}{s}L\left(2u^2u_x - u_x^2 - \frac{1}{2}u_{xx}u\right)\right). \tag{5.419}$$

Apply the convex homotopy

$$\begin{aligned} &u_0 + pu_1 + p^2u_2 + \cdots \\ &= \left(-\frac{1}{2} + \frac{1}{1+e^x}\right) - \int_0^t L^{-1}\left(\frac{1}{s}L\left(2(u_0 + pu_1 + \cdots)^2\left(\frac{\partial u_0}{\partial x} + p\frac{\partial u_1}{\partial x} + \cdots\right)\right)\right)dt \\ &+ \int_0^t L^{-1}\left(\frac{1}{s}L(u_0 + pu_1 + \cdots)_x^2 + \frac{1}{2}\left(\frac{\partial^2 u_0}{\partial x^2} + p\frac{\partial^2 u_1}{\partial x^2} + \cdots\right)(u_0 + pu_1 + \cdots)\right)dt. \end{aligned} \tag{5.420}$$

Comparing the coefficient of like powers of p,

$$\begin{aligned} p^{(0)} &: u_0(x,t) = \left(-\frac{1}{2} + \frac{1}{1+e^x}\right), \\ p^{(1)} &: u_1(x,t) = \frac{4e^x t}{(1+2e^x)^4} + 8\frac{e^{2x}t}{(1+2e^x)^4} - \frac{5e^x t}{(1+2e^x)^3} - \frac{2e^{2x}t}{(1+2e^x)^3} \\ &\quad + \frac{3e^x t}{2(1+2e^x)^2} - \frac{e^x t^2}{16(1+2e^x)^3} + \frac{e^{2x}t}{8(1+2e^x)^3}. \\ &\vdots \end{aligned} \tag{5.421}$$

The series solution is given by

$$\begin{aligned} u(x,t) = &\left(-\frac{1}{2} + \frac{1}{1+e^x}\right) + \frac{4e^x t}{(1+2e^x)^4} + 8\frac{e^{2x}t}{(1+2e^x)^4} - \frac{5e^x t}{(1+2e^x)^3} \\ &- \frac{2e^{2x}t}{(1+2e^x)^3} + \frac{3e^x t}{2(1+2e^x)^2} - \frac{e^x t^2}{16(1+2e^x)^3} + \frac{e^{2x}t}{8(1+2e^x)^3}\cdots. \end{aligned} \tag{5.422}$$

Applications of the homotopy perturbation method to solitary solutions can be referred to [74–77].

5.8. Parameter-Expansion Method

5.8.1. An Example

Slab detachment or break-off is appreciated as an important geological process. An analytical solution was given in [78] for the nonlinear dynamics of high amplitude necking in a free layer of power-law fluid extended in layer-parallel direction due to buoyancy stress.

The power-law flow law for the layer can be written as

$$\dot{\varepsilon} = B\tau^n, \tag{5.423}$$

where $\dot{\varepsilon}$, rate of deformation parallel to the layer of a plane in the layer, τ is the mean layer-parallel deviatoric stress, and B and n are a material parameter and the stress exponent, respectively.

Schmalholz obtained the following detachment time [78]:

$$t = \frac{t_c}{n} = \frac{1}{nB((1/2)F)^n}, \tag{5.424}$$

where F is the layer-parallel force due to buoyancy at the top of the layer.

The power-law fluid assumption, (5.423), is useful because of its simplicity but only approximately describes the behaviour of a real non-Newtonian fluid. We modify (5.423) in the form

$$\dot{\varepsilon} = B_1\tau + B_2\tau^n, \tag{5.425}$$

where B_1 and B_2 are material parameters.

The governing equation for the dynamics of necking becomes

$$-\frac{1}{D}\frac{dD}{dt} = \frac{B_1 F}{2D} + B_2\left(\frac{F}{2D}\right)^n, \qquad D(0) = D_0, \tag{5.426}$$

where D is the local thickness of the layer, and F is the layer-parallel force.

Rewrite (5.426) in the form

$$\frac{dD}{dt} + A_1 + A_2 D^{1-n} = 0, \qquad D(0) = D_0, \tag{5.427}$$

where $A_1 = B_1F/2$, $A_2 = B_2(F/2)^n$.

We use the parameter-expansion method (see Section 5.8.2) to solve (5.427). We expand the solution in the form

$$D = \overline{D} + p\widetilde{D} + p^2\widetilde{\widetilde{D}} + \cdots, \tag{5.428}$$

where $\overline{D}$ is a reference thickness, $\widetilde{D}$ is a perturbation term, $\widetilde{\widetilde{D}}$ is an even smaller infinitesimal perturbation term, and p is a bookkeeping parameter, $p = 1$ (see Section 5.8.2).

The parameters A_1 and A_2 in (5.427) are expended in a similar way:

$$\begin{aligned} A_1 &= \overline{A}_1 + p\widetilde{A}_1 + p^2\widetilde{\widetilde{A}}_1 + \cdots, \\ A_2 &= p\left(\overline{A}_2 + p\widetilde{A}_2 + p^2\widetilde{\widetilde{A}}_2 + \cdots\right). \end{aligned} \tag{5.429}$$

Substituting (5.428)-(5.429) into (5.427), collecting terms of the same power of p, and equating coefficients of like powers of p yield a series of linear equations. We write only the equations for $\overline{D}$ and $\widetilde{D}$, which read

$$\frac{d\overline{D}}{dt} + \overline{A}_1 = 0, \qquad \overline{D}(0) = D_0, \tag{5.430}$$

$$\frac{d\widetilde{D}}{dt} + \widetilde{A}_1 + \overline{A}_2\overline{D}^{1-n} = 0, \qquad \widetilde{D}(0) = 0. \tag{5.431}$$

The solution of (5.430) reads

$$\overline{D} = D_0 - \overline{A}_1 t. \tag{5.432}$$

Substituting the result into (5.431), we have

$$\frac{d\widetilde{D}}{dt} + \widetilde{A}_1 + \overline{A}_2\left(D_0 - \overline{A}_1 t\right)^{1-n} = 0, \qquad \widetilde{D}(0) = 0. \tag{5.433}$$

Its solution is

$$\widetilde{D} = -\frac{\overline{A}_2(D_0)^{2-n}}{(2-n)\overline{A}_1} - \widetilde{A}_1 t + \frac{\overline{A}_2\left(D_0 - \overline{A}_1 t\right)^{2-n}}{(2-n)\overline{A}_1}. \tag{5.434}$$

Note that $\widetilde{D}$ is a perturbation term. When t is small, it can be approximated as

$$\begin{aligned} \widetilde{D} &= -\frac{\overline{A}_2(D_0)^{2-n}}{(2-n)\overline{A}_1} - \widetilde{A}_1 t + \frac{\overline{A}_2}{(2-n)\overline{A}_1}\left(D_0^{2-n} - (2-n)D_0^{1-n}\overline{A}_1 t\right) \\ &= -\left(\widetilde{A}_1 - \overline{A}_2 D_0^{1-n}\right)t. \end{aligned} \tag{5.435}$$

We set

$$\widetilde{A}_1 = \overline{A}_2 D_0^{1-n}. \tag{5.436}$$

If the first-order approximate solution is enough, setting $p = 1$ in (5.428)-(5.429) and equating higher-order perturbation terms to be zero, we have

$$
\begin{aligned}
A_2 &= \bar{A}_2, \\
\tilde{A}_1 &= A_2 D_0^{1-n}, \\
\bar{A}_1 &= A_1 - A_2 D_0^{1-n}, \\
D = \bar{D} + \tilde{D} &= D_0 - \bar{A}_1 t + \frac{A_2 (D_0)^{2-n}}{(2-n)\bar{A}_1} - \tilde{A}_1 t - \frac{A_2\left(D_0 - \bar{A}_1 t\right)^{2-n}}{(2-n)\bar{A}_1}.
\end{aligned}
\tag{5.437}
$$

For qualitative analysis, we use $\bar{D}$ as an approximate solution

$$
D = D_0 - \left(A_1 - A_2 D_0^{1-n}\right) t. \tag{5.438}
$$

The detachment time is

$$
t = \frac{D_0}{A_1 - A_2 D_0^{1-n}} = \frac{D_0}{(B_1 F/2) - B_2 (F/2)^n D_0^{1-n}}. \tag{5.439}
$$

Equation (5.439) shows that the detachment time depends upon the initial layer thickness and fluid properties and the buoyancy.

The nonlinear equation describing the dynamics of necking is of strong nonlinearity, and in order to solve it analytically, oversimple assumption has to be made. The power-law fluid is simple, but it is not a real non-Newtonian fluid. Recent development of analytical methods increases tantalizing possibility of theoretically seeking the approximate solutions to the present problem.

5.8.2. Parameter-Expansion Method for Solitary Solutions

Parameter-expansion method includes the modified Lindstedt-Poincare method [79] and bookkeeping parameter method [80]. In the review article [7], it was previously called parameter-expanding method. The method does not require to construct a homotopy. To illustrate its solution procedure, we reconsider (5.384), which is rewritten in the form

$$
\frac{\partial u}{\partial t} + a u \frac{\partial u}{\partial x} + b \frac{\partial^3 u}{\partial x^3} + 1 \cdot N(u) = 0. \tag{5.440}
$$

Supposing that the parameters, a, b, and 1, can be expressed in the forms

$$
\begin{aligned}
a &= a_0 + p a_1 + p^2 a_2 + \cdots, \\
b &= b_0 + p b_1 + p^2 b_2 + \cdots, \\
1 &= p c_1 + p^2 c_2 + \cdots,
\end{aligned}
\tag{5.441}
$$

where p is a bookkeeping parameter, $p = 1$, substituting (5.441) into (5.440), proceeding the same way as the perturbation method, we can easily obtain the needed solution.

We consider a simple mathematical model in the form

$$u'' + u^2 = 0, \qquad u(0) = u(1) = 0,$$
$$u'' + 0 + 1 \cdot u^2 = 0. \tag{5.442}$$

We seek an expansion of the form

$$u = u_0 + pu_1 + p^2 u_2 + \cdots, \tag{5.443}$$

where the ellipsis dots stand for terms proportional to powers of p greater than 2, p is a bookkeeping parameter, $p = 1$.

The constants, 0 and 1, in the left-hand side of (5.442) can be, respectively, expanded in a similar way

$$0 = 2a + a_1 p + a_2 p^2 + \cdots,$$
$$1 = b_1 p + b_2 p^2 + \cdots. \tag{5.444}$$

Substituting (5.444) to (5.442), we have

$$\left(u_0 + pu_1 + p^2 u_2 + \cdots\right)'' + \left(2a + a_1 p + a_2 p^2 + \cdots\right)$$
$$+ \left(b_1 p + b_2 p^2 + \cdots\right) \cdot \left(u_0 + pu_1 + p^2 u_2 + \cdots\right)^2 = 0, \tag{5.445}$$

and equating coefficients of like powers of p we obtain the same equations as illustrated in previous examples:

$$u_0'' + 2a = 0, \qquad u_0(0) = u_0(1) = 0,$$
$$u_1'' + b_1 u_0^2 + a_1 = 0, \qquad u_1(0) = u_1(1) = 0. \tag{5.446}$$

Solving u_0 and u_1, we have the following first-order approximate solution:

$$u(t) = u_0(t) + u_1(t) = at(1-t) + at^2 - a^2\left(\frac{1}{30}t^6 - \frac{1}{10}t^5 + \frac{1}{12}t^4\right) - \left(a - \frac{1}{60}a^2\right)t. \tag{5.447}$$

The discussion on how to determine the constant, a, is given in [69]. Applications of the parameter-expansion method to various nonlinear problems are available in [81–86].

5.9. Exp-Function Method

5.9.1. Exp-Function Method for Solitary Solutions

Exp-function method [87–89] provides us with a straightforward and concise approach for obtaining generalized solitary solutions and periodic solutions; the solution procedure, by the help of Matlab or Mathematica, is of utter simplicity, see [90–97]. Considering a general nonlinear partial differential equation in the form

$$F(u,u_x,u_y,u_z,u_t,u_{xx},u_{yy},u_{zz},u_{tt},u_{xy},u_{xt},u_{yt},\ldots)=0. \tag{5.448}$$

The exp-function method assumes that the solitary solution can be expressed in the form

$$u(x,y,z,t)=\frac{\sum_{n=-k}^{l} p_n \exp(a_n x+b_n y+c_n z+d_n t)}{\sum_{m=-i}^{j} q_m \exp(a_m x+b_m y+c_m z+d_m t)}. \tag{5.449}$$

We can also introduce a transformation

$$\xi=ax+by+cz+dt. \tag{5.450}$$

We can rewrite (5.448) in the following nonlinear ordinary differential equation:

$$F(u,u',u'',u''',\ldots)=0, \tag{5.451}$$

where the prime denotes the derivation with respect to ξ.

According to the exp-function method, the traveling wave solutions can be expressed in the form

$$u(\xi)=\frac{\sum_{n=-k}^{l} a_n \exp(n\xi)}{\sum_{m=-i}^{j} b_m \exp(m\xi)}, \tag{5.452}$$

where i, j, k, and l are positive integers which could be freely chosen; a_n and b_m are unknown constants to be determined.

Consider a fractional Benjamin-Bona-Mahony (BBM) equation [97]

$$\frac{\partial u}{\partial t}-u\frac{\partial u}{\partial x}-\frac{\partial}{\partial t}\frac{\partial^2 u}{\partial x^2}=0, \tag{5.453}$$

using a transformation

$$u(x,t)=U(\xi),\quad \xi=kx+wt, \tag{5.454}$$

(5.453) becomes an ordinary differential equation,

$$wU'-kUU'-k^2wU'''=0. \tag{5.455}$$

By the exp-function method, we assume that the solution of (5.460) can be expressed as

$$U(\xi) = \frac{a_1 \exp(\xi) + a_0 + a_{-1} \exp(-\xi)}{b_1 \exp(\xi) + b_0 + b_{-1} \exp(-\xi)}, \tag{5.456}$$

where $a_i, b_i (i = 1, 0, -1)$ are unknown parameters.

Substituting (5.456) into (5.455), collecting terms of the same term of $\exp(i\xi)$, we have

$$\frac{1}{A}[C_7 \exp(7\xi) + C_6 \exp(6\xi) + C_5 \exp(5\xi) + C_4 \exp(4\xi) + C_3 \exp(3\xi) + C_2 \exp(2\xi) + C_1 \exp(\xi)] = 0. \tag{5.457}$$

Equating coefficients, $C_i (i = 1 \sim 7)$, to zero yields a series of linear equations

$$C_i = 0 \quad (i = 1 \sim 7). \tag{5.458}$$

Solving the system of algebraic equation, we can identify parameters in (5.456). Finally, we obtain [97]

$$u(x,t) = -\frac{w(k^2 - 1)}{k} + \frac{6k^2 b_0}{b_1 \exp[(kx + wt)] + b_0 + (b_0^2/4b_1) \exp[-(kx + wt)]}, \tag{5.459}$$

where b_0 and b_1 are free parameters.

By $k = iK$, $w = iW$, the obtained solitary solution can be converted into periodic solution.

Other expressions are listed as follows:

$$u(x,t) = \frac{\sum_n a_n \exp(q_n x + p_n t)}{\sum_m b_m \exp(\alpha_m x + \beta_m t)}, \tag{5.460}$$

where $a_i, b_i \alpha_i, \beta_i q_i$, and p_i are unknown constants to be further determined.

For a two-wave solution, we can use the double exp-function method [8], which requires

$$u = \frac{a_{-1} \exp(-\xi) + b_1 \exp(\xi) + a_0 + b_{-1} \exp(-\eta) + b_1 \exp(\eta)}{c_{-1} \exp(-\xi) + c_1 \exp(\xi) + b_0 + d_{-1} \exp(-\eta) + d_1 \exp(\eta)}, \tag{5.461}$$

where $\xi = a_1 x + b_1 y + c_1 z - d_1 t$ and $\eta = a_2 x + b_2 y + c_2 z - d_2 t$.

5.9.2. The Exp-Function Method for Periodic Solutions

The French mathematician J. Fourier (1768–1830) showed that any periodic motion can be represented by a series of sines and cosines that are harmonically related

$$u(t) = \frac{1}{2}a_0 + a_1 \cos \omega_1 t + a_2 \cos \omega_2 t + \cdots + b_1 \sin \omega_1 t + b_2 \sin \omega_2 t + \cdots, \tag{5.462}$$

where $\omega_n = n\omega_1$, and a_n and b_n are, respectively, expressed as

$$\begin{aligned} a_n &= \frac{2}{T}\int_{-T/2}^{T/2} u(t) \cos \omega_n t \, dt, \\ b_n &= \frac{2}{T}\int_{-T/2}^{T/2} u(t) \sin \omega_n t \, dt. \end{aligned} \tag{5.463}$$

Using Euler's formula, we have

$$\begin{aligned} \sin t &= \frac{1}{2}\left(e^{it} + e^{-it}\right), \\ \cos t &= -\frac{1}{2}i\left(e^{it} - e^{-it}\right). \end{aligned} \tag{5.464}$$

We, therefore, can rewrite the Fourier series, (5.462), in the form

$$u(t) = \sum_{n=-\infty}^{\infty} c_n e^{i\omega_n t}, \tag{5.465}$$

where $c_0 = (1/2)a_0$ and $c_n = (1/2)(a_n - ib_n)$.

For periodic solution, we can assume that the solution can be expressed in the form

$$u(t) = \frac{\sum_{n=-c}^{d} p_n \exp(i\omega_n t)}{\sum_{m=-f}^{g} q_m \exp(i\omega_m t)}, \tag{5.466}$$

or

$$u(t) = \sum_{0}^{N} c_n e^{i\omega_n t}. \tag{5.467}$$

5.10. Subequation Method

There are many subequation methods in the open literature for solitary solution [98–100]. We consider a general nonlinear equation in the form

$$F(u, u_t, u_x, u_y, u_z, u_{xx}, u_{yy}, u_{zz}, \ldots) = 0. \tag{5.468}$$

The sub-equation method follows the following steps.

Step 1. Let $u = u(x,y,z,t) = u(\xi)$, where $\xi = ax + by + cz + dt + \xi_0$. Equation (5.468) turns out to be an ordinary differential equation

$$F(u, u_\xi, u_{\xi\xi}, u_{\xi\xi\xi}, \ldots) = 0. \tag{5.469}$$

Step 2. Assume that the solution can be expressed in the form

$$u = \sum_n a_n f^n(\xi), \tag{5.470}$$

or in a more general form,

$$u = U(f(\xi)), \tag{5.471}$$

where f is the solution of an auxiliary equation

$$f'^2 = b_0 + b_1 f + b_2 f^2 + b_3 f^3 + b_4 f^4 + \cdots. \tag{5.472}$$

The exact solution of (5.472) for fixed $b_i (i = 0, 1, 2, 3, \ldots)$ is known.

Step 3. Use some mathematical software to determine a_n in (5.470).

The auxiliary equation can be constructed using the Hamiltonian invariant as discussed in Section 5.3.5. For (5.128), and the following auxiliary equation can be constructed:

$$u' = \sqrt{cu^2 + 2u^3}. \tag{5.473}$$

For (5.227), the following auxiliary equation should be in the form:

$$u' = \sqrt{cu^2 - \frac{12}{(n+1)(n+2)} u^{n+2}}. \tag{5.474}$$

In such case, the solution of the auxiliary equation is the special solution of the problem.

5.11. Ancient Chinese Mathematics

5.11.1. Introduction to the Ancient Chinese Mathematics

Can ancient Chinese mathematics still work for nonlinear science? Yes, it can. It is simple but effective.

To illustrate the basic idea of the ancient Chinese method, we consider an algebraic equation:

$$f(x) = 0. \tag{5.475}$$

Let x_1 and x_2 be the approximate solutions of the equation, which lead to the remainders $f(x_1)$ and $f(x_2)$, respectively; the ancient Chinese algorithm leads to the result

$$x = \frac{x_2 f(x_1) - x_1 f(x_2)}{f(x_1) - f(x_2)}. \tag{5.476}$$

Now we consider a generalized nonlinear oscillator in the form

$$u'' + f(u) = 0, \qquad u(0) = A, \qquad u'(0) = 0. \tag{5.477}$$

We use two trial functions $u_1(t) = A\cos\omega_1 t$ and $u_2 = A\cos\omega_2 t$, which are, respectively, the solutions of the following linear oscillator equations:

$$\begin{aligned} u'' + \omega_1^2 u &= 0, \\ u'' + \omega_2^2 u &= 0. \end{aligned} \tag{5.478}$$

The residuals are

$$\begin{aligned} R_1(\omega_1 t) &= -\omega_1^2 \cos\omega_1 t + f(A\cos\omega_1 t), \\ R_2(\omega_2 t) &= -\omega_2^2 \cos\omega_2 t + f(A\cos\omega_2 t). \end{aligned} \tag{5.479}$$

The residuals depend upon t, and in our previous applications [7], we just use $R_1(0)$ and $R_2(0)$ for simplicity when we use the ancient Chinese method, resulting in relatively low accuracy [7]

$$\omega^2 = \frac{\omega_1^2 R_2(0) - \omega_2^2 R_1(0)}{R_2(0) - R_1(0)}. \tag{5.480}$$

Equation (5.480) is called He's frequency formulation or He's frequency-amplitude formulation in the literature [101–106].

In order to improve its accuracy, we introduce two new residual variables $\tilde{R}_1$ and $\tilde{R}_2$ defined as [105, 106]

$$\tilde{R}_1 = \frac{4}{T_1}\int_0^{T_1/4} R_1(t)\cos\left(\frac{2\pi}{T_1}t\right)dt, \tag{5.481}$$

$$\tilde{R}_2 = \frac{4}{T_2}\int_0^{T_2/4} R_2(t)\cos\left(\frac{2\pi}{T_2}t\right)dt. \tag{5.482}$$

According to the ancient Chinese method, we can approximately determine ω^2 in the form

$$\omega^2 = \frac{\omega_1^2 \widetilde{R}_2 - \omega_2^2 \widetilde{R}_1}{\widetilde{R}_2 - \widetilde{R}_1}. \tag{5.483}$$

In this letter, we will use the well-known Duffing equation.

Using trial functions $u_1(t) = A\cos t$ and $u_2 = A\cos\omega_2 t$, respectively, for (5.481), we obtain the following residuals:

$$\begin{aligned} R_1(t) &= \varepsilon A^3 \cos^3 t, \\ R_2(t) &= A\left(1 - \omega_2^2\right)\cos\omega_2 t + \varepsilon A^3 \cos^3 \omega_2 t, \end{aligned} \tag{5.484}$$

where $\omega_2 > 0$ and $\omega_2 \neq 1$.

By simple calculation, we obtain

$$\begin{aligned} \widetilde{R}_1 &= \frac{4}{T_1}\int_0^{T_1/4} R_1(t)\cos\left(\frac{2\pi}{T_1}t\right)dt = \frac{2}{\pi}\int_0^{\pi/2} \varepsilon A^3 \cos^4 t\,dt = \frac{3}{4\pi}, \\ \widetilde{R}_2 &= \frac{4}{T_2}\int_0^{T_2/4} R_2(t)\cos\left(\frac{2\pi}{T_2}t\right)dt = \frac{2\omega}{\pi}\int_0^{T_2/4}\left\{A\left(1-\omega_2^2\right)\cos^2\omega_2 t + \varepsilon A^3\cos^4\omega_2 t\right\}dt \\ &= \frac{2}{\pi}\int_0^{\pi/2}\left\{A\left(1-\omega_2^2\right)\cos^2 s + \varepsilon A^3 \cos^4 s\right\}ds \\ &= A\left(1-\omega_2^2\right)\frac{1}{\pi} + \frac{3}{4\pi}\varepsilon A^3. \end{aligned} \tag{5.485}$$

Applying (5.483), we have

$$\begin{aligned} \omega^2 &= \frac{\omega_1^2 \widetilde{R}_2 - \omega_2^2 \widetilde{R}_1}{\widetilde{R}_2 - \widetilde{R}_1} \\ &= \frac{A(1-\omega_2^2)(1/\pi) + (3/4\pi)\varepsilon A^3 - (3/4\pi)\varepsilon A^3 \omega_2^2}{A(1-\omega_2^2)(1/\pi) + (3/4\pi)\varepsilon A^3 - (3/4\pi)\varepsilon A^3} \\ &= \frac{(1-\omega_2^2) + (3/4)\varepsilon A^3 - (3/4)\varepsilon A^3\omega_2^2}{(1-\omega_2^2)} \\ &= 1 + \frac{3}{4}\varepsilon A^2. \end{aligned} \tag{5.486}$$

5.11.2. *Ancient Chinese Mathematics for Optimization [107]*

To elucidate the basic idea of application of the ancient Chinese mathematics to optimal problems, we consider first the following simple example:

$$f = xy \longrightarrow \max, \tag{5.487}$$

$$\text{s.t.} \quad g(x,y) = x + y - 1 = 0. \tag{5.488}$$

To make the Lagrange function f maximal, it requires

$$df = xdy + ydx = 0, \tag{5.489}$$

where dx and dy are not independent, and according to the constraint equation, (5.490), we have

$$dx + dy = 0. \tag{5.490}$$

Considering the relationship between dx and dy, (5.492), (5.491) becomes

$$df = (x - y)dy = 0. \tag{5.491}$$

Due to the arbitrary dy in (5.491), we have the following Euler equation:

$$x - y = 0. \tag{5.492}$$

Solving (5.488) and (5.492) simultaneously, we find the optimal values

$$x^* = y^* = \frac{1}{2}. \tag{5.493}$$

A general solution procedure is given in [8], and the stationary condition (Euler equation) can be simply obtained using He's brackets [8, 108]:

$$\langle f, g \rangle_{x,y} = f_x g_y - f_y g_x = 0, \tag{5.494}$$

where $f_x = \partial f/\partial x = y$, $f_y = \partial f/\partial y = x$, $g_x = \partial g/\partial x = 1$, and $g_y = \partial g/\partial y = 1$.

Equation (5.494) is exactly the same with (5.492).

According to the ancient Chinese mathematics, to solve a problem, we should choose double trial solutions, x_1 and x_2. We can choose x_1 arbitrarily, for example, $x_1 = 2/7$, according to the constraint equation (5.488), and we have $y_1 = 5/7$.

From (5.487), we obtain

$$f_1 = x_1 y_1 = \frac{10}{49}. \tag{5.495}$$

The condition (5.489) can be approximately expressed as

$$\Delta f = x\Delta y + y\Delta x = 0, \tag{5.496}$$

where $\Delta x = x_2 - x_1$, $\Delta y = y_2 - y_1$, and $\Delta f = f_2 - f_1$.

Equation (5.496) implies that

$$f_2 = f_1. \tag{5.497}$$

This means that

$$f_2 = x_2 y_2 = \frac{10}{49}. \tag{5.498}$$

Considering the constraint equation, (5.488), again, we have

$$x_2 + y_2 = 1. \tag{5.499}$$

From (5.498) and (5.499), we can obtain immediately $x_2 = 5/7$ and $y_2 = 2/7$.

Substituting x_1, x_2, y_1, and y_2 into (5.498), we have

$$\Delta f = x(y_2 - y_1) + y(x_2 - x_1) = x\left(\frac{2}{7} - \frac{5}{7}\right) + y\left(\frac{5}{7} - \frac{2}{7}\right) = 0. \tag{5.500}$$

Solving (5.488) and (5.500) simultaneously, we find the optimal values given in (5.493).

Example 5.1. Condiser the following [109]:

$$f = \sqrt{x^2 + y^2} \longrightarrow \min, \tag{5.501}$$

$$\text{s.t.} \quad g(x, y) = y - 2 - x^2 = 0. \tag{5.502}$$

To make (5.501) minimal, it requires that

$$\Delta f = \frac{2x\Delta x + 2y\Delta y}{2\sqrt{x^2 + y^2}} = \frac{x\Delta x + y\Delta y}{\sqrt{x^2 + y^2}} = 0. \tag{5.503}$$

Choosing $x_1 = 1$, we have $y_1 = 3$ and $f_1 = \sqrt{10}$. We set $f_2 = \sqrt{10}$, which means that

$$x_2^2 + y_2^2 = 10, \tag{5.504}$$

$$y_2 = 2 + x_2^2. \tag{5.505}$$

Solving (5.504) and (5.505) simultaneously, we have $x_2 = -1$ and $y_2 = 3$.

Substituting x_1, x_2, y_1, and y_2 into (5.503), we obtain

$$\Delta f = \frac{x(x_2 - x_1) + y(y_2 - y_1)}{\sqrt{x^2 + y^2}} = \frac{-2x}{\sqrt{x^2 + y^2}} = 0. \tag{5.506}$$

We, therefore, obtain the optimal values $x^* = 0$ and $y^* = 2$.

Example 5.2. Consider the following problem [109]:

$$f(x, y) = \sqrt{x^2 + y^2} \longrightarrow \min, \tag{5.507}$$

$$\text{s.t.} \quad y^2 - (x-1)^3 = 0. \tag{5.508}$$

The Lagrange multiplier method becomes invalid for this example [109]. Using the basic idea of the ancient Chinese mathematics, we choose $x_1 = 2$, $y_1 = 1$ and $x_2 = 2$, $y_2 = -1$, and both cases lead to $f_1 = f_2 = \sqrt{5}$.

Making (5.507) stationary requires

$$\Delta f = \frac{x\Delta x + y\Delta y}{\sqrt{x^2 + y^2}} = \frac{x(x_2 - x_1) + y(y_2 - y_1)}{\sqrt{x^2 + y^2}} = \frac{-2y}{\sqrt{x^2 + y^2}} = 0, \tag{5.509}$$

which leads to immediately $y^* = 0$ and $x^* = 1$ if the constraint equation, (5.508), is considered.

5.11.3. *Ancient Chinese Mathematics for Nonlinear Wave Equations [110]*

Now we consider the nonlinear Schrödinger equation

$$iu_t + u_{xx} + 2|u|^2 u = 0, \qquad u(x, 0) = e^{ikx}. \tag{5.510}$$

We use the trial functions

$$u_1(x, t) = e^{i(kx+t)}, \quad u_2(x, t) = e^{i(kx+\omega t)}. \tag{5.511}$$

The residuals are

$$R_1(x, t) = \left(1 - k^2\right) e^{i(kx+t)}, \quad R_2(x, t) = \left(2 - \omega - k^2\right) e^{i(kx+\omega t)}. \tag{5.512}$$

According to frequency-amplitude formulation, (5.480), we have

$$\omega^2 = \frac{\omega_1^2 R_2(x,0) - \omega_2^2 R_1(x,0)}{R_2(x,0) - R_1(x,0)} = -\left(\omega + \omega k^2 - 2\omega + k^2 - 2\right). \tag{5.513}$$

We get the equation about ω, and the solution is

$$\begin{aligned} \omega &= 2 - k^2, \\ u(x,t) &= e^{i\{kx+(2-k^2)\omega t\}}. \end{aligned} \tag{5.514}$$

Now we consider the Schrödinger equation with power law nonlinearity

$$iu_t + u_{xx} - 2|u|^{2r}u = 0, \quad u(x,0) = \left(2(r+1)\operatorname{sech}^2(2rx)\right)^{1/2r}, \quad r \geq 1. \tag{5.515}$$

Following the analysis presented above, we choose the trial functions

$$\begin{aligned} u_1(x,t) &= \left(2(r+1)\operatorname{sech}^2(2rx)\right)^{1/2r} e^{it}, \\ u_2(x,t) &= \left(2(r+1)\operatorname{sech}^2(2rx)\right)^{1/2r} e^{i\omega t}. \end{aligned} \tag{5.516}$$

The residuals are

$$\begin{aligned} R_1(x,t) = &-\left(2(r+1)(\operatorname{sech}(2rx))^2\right)^{1/2r} e^{it} + 4\left(2(r+1)(\operatorname{sech}(2rx))^2\right)^{1/2r}(\tanh(2rx))^2 e^{it} \\ &- 4\left(2(r+1)(\operatorname{sech}(2rx))^2\right)^{1/2r}\left(1 - (\tanh(2rx))^2\right) r e^{it} \\ &+ 2\left(2(r+1)(\operatorname{sech}(2rx))^2\right)^{1/2r^{-1}} e^{it}, \\ R_2(x,t) = &-\left(2(r+1)(\operatorname{sech}(2rx))^2\right)^{1/2r} e^{i\omega t}\omega + 4\left(2(r+1)(\operatorname{sech}(2rx))^2\right)^{1/2r}(\tanh(2rx))^2 e^{i\omega t} \\ &- 4\left(2(r+1)(\operatorname{sech}(2rx))^2\right)^{1/2r}\left(1 - (\tanh(2rx))^2\right) r e^{i\omega t} \\ &+ 2\left(2(r+1)(\operatorname{sech}(2rx))^2\right)^{1/2r} e^{i\omega t}. \end{aligned} \tag{5.517}$$

According to the frequency-amplitude formulation, (5.480), setting $x = 0$ and $t = 0$, we have

$$\omega^2 = -4r + 2\omega\left|\left(2^{1/2r}(r+1)^{1/2r}\right)^{2r}\right| - \omega - 4\omega r + \left|\left(2^{1/2r}(r+1)^{1/2r}\right)^{2r}\right|. \tag{5.518}$$

Solving ω from the above equation yields

$$\omega = -4r + 2\left|\left(2^{1/2r}(r+1)^{1/2r}\right)^{2r}\right|. \tag{5.519}$$

We, therefore, obtain the solution

$$u_2(x,t) = \left(2(r+1)\operatorname{sech}^2(2rx)\right)^{1/2r} e^{i\omega t}. \tag{5.520}$$

5.11.4. Solitary-Solution Formulation for Nonlinear Equations

Now consider a general nonlinear equation

$$L(u) + N(u) = 0, \tag{5.521}$$

where L is a linear operator, and N is a nonlinear operator.

We choose two trial equations

$$\begin{aligned} L(u) &= 0, \\ \widetilde{L}(u) &= 0, \end{aligned} \tag{5.522}$$

where $\widetilde{L}$ is a linearized linear operator or a simple nonlinear operator, and its solution $\widetilde{L}(u) = 0$ is known. For example, we consider a nonlinear oscillator: $u'' + u^3 = 0$, and we can choose $L(u) = u'' + u = 0$ and $\widetilde{L}(u) = u'' + \omega^2 u = 0$, where ω is the frequency to be searched for.

Substituting the solutions of (5.522) with same boundary/initial conditions as (5.521) into (5.521), we obtain, respectively, the residuals R_1 and R_2.

For traveling wave solutions, we can make a transformation

$$\xi = ax + by + cz + \omega t. \tag{5.523}$$

The solitary-solution formulation can be expressed as

$$\omega^2 = \frac{R_1\omega_1^2 - R_2\omega_2^2}{R_1 - R_2}, \tag{5.524}$$

or

$$\omega^n = \frac{R_1\omega_1^n - R_2\omega_2^n}{R_1 - R_2}. \tag{5.525}$$

The KdV equation is a nonlinear, dispersive partial differential equation for a function of two real variables, space x and time t:

$$u_t + 6uu_x + u_{xxx} = 0. \tag{5.526}$$

According to the solitary-solution formulation (5.524), we construct two trial functions in the following forms:

$$
\begin{aligned}
u_1(x,t) &= \frac{A}{\exp(kx+t+D)+\exp[-(kx+t+D)]+B'},\\
u_2(x,t) &= \frac{A}{\exp(kx+\omega t+D)+\exp[-(kx+\omega t+D)]+B'},
\end{aligned}
\tag{5.527}
$$

where A, B, D are arbitrary constants.

Substituting u_1 and u_2 into (5.526), respectively, we obtain the following residuals:

$$
\begin{aligned}
&R_1(x,t)\\
&= \frac{-A\{\exp(kx+t+D)-\exp[-(kx+t+D)]\}}{\{\exp(kx+t+D)+\exp[-(kx+t+D)]+B\}^2}\\
&\quad - \frac{6A^2k\{\exp(kx+t+D)-\exp[-(kx+t+D)]\}}{\{\exp(kx+t+D)+\exp[-(kx+t+D)]+B\}^3}\\
&\quad - \frac{Ak^3\{\exp(kx+t+D)-\exp[-(kx+t+D)]\}}{\{\exp(kx+t+D)+\exp[-(kx+t+D)]+B\}^2}\\
&\quad - \frac{6Ak^3\{\exp(kx+t+D)-\exp[-(kx+t+D)]\}^3}{\{\exp(kx+t+D)+\exp[-(kx+t+D)]+B\}^4}\\
&\quad + \frac{6Ak^3\{\exp(kx+t+D)-\exp[-(kx+t+D)]\}\{\exp(kx+t+D)+\exp[-(kx+t+D)]\}}{\{\exp(kx+t+D)+\exp[-(kx+t+D)]+B\}^3},\\
&R_2(x,t)\\
&= \frac{-A\omega\{\exp(kx+\omega t+D)-\exp[-(kx+\omega t+D)]\}}{\{\exp(kx+\omega t+D)+\exp[-(kx+\omega t+D)]+B\}^2}\\
&\quad - \frac{6A^2k\{\exp(kx+\omega t+D)-\exp[-(kx+\omega t+D)]\}}{\{\exp(kx+\omega t+D)+\exp[-(kx+\omega t+D)]+B\}^3}\\
&\quad - \frac{Ak^3\{\exp(kx+\omega t+D)-\exp[-(kx+\omega t+D)]\}}{\{\exp(kx+\omega t+D)+\exp[-(kx+\omega t+D)]+B\}^2}\\
&\quad - \frac{6Ak^3\{\exp(kx+\omega t+D)-\exp[-(kx+\omega t+D)]\}^3}{\{\exp(kx+\omega t+D)+\exp[-(kx+\omega t+D)]+B\}^4}\\
&\quad + \frac{6Ak^3\{\exp(kx+\omega t+D)-\exp[-(kx+\omega t+D)]\}\{\exp(kx+\omega t+D)+\exp[-(kx+\omega t+D)]\}}{\{\exp(kx+\omega t+D)+\exp[-(kx+\omega t+D)]+B\}^3}.
\end{aligned}
\tag{5.528}
$$

Using the solitary-solution formulation (5.480), and choosing the location point $(0,0)$, we obtain

$$\omega^2 = \frac{\omega_1^2 R_2(0,0) - \omega_2^2 R_1(0,0)}{R_2(0,0) - R_1(0,0)}. \tag{5.529}$$

Simplifying (5.529) results in

$$\begin{aligned} &\omega^2 + \omega + (\omega+1)k^3 + \left[2B\omega^2 + 2\omega B + (\omega+1)\left(6kA - 4k^3B\right)\right]\exp(-D) \\ &+ \left[2\omega^2 + B^2\omega^2 + 2\omega + \omega B^2 + (\omega+1)\left(6kAB - 22k^3 + k^3B^2\right)\right]\exp(-2D) \\ &+ \left[2B\omega^2 + 2\omega B + (\omega+1)\left(6kA - 4k^3B\right)\right]\exp(-3D) \\ &+ \left[\omega^2 + \omega + (\omega+1)k^3\right]\exp(-4D) = 0. \end{aligned} \tag{5.530}$$

Setting the coefficients of $\exp(-nD)$ $(n = 0,1,2,3,4)$ to be zero, we obtain

$$\omega^2 + \omega + (\omega+1)k^3 = 0, \tag{5.531}$$

$$2B\omega^2 + 2\omega B + (\omega+1)\left(6kA - 4k^3B\right) = 0, \tag{5.532}$$

$$2\omega^2 + B^2\omega^2 + 2\omega + \omega B^2 + (\omega+1)\left(6kAB - 22k^3 + k^3B^2\right) = 0. \tag{5.533}$$

Solving (5.531)~(5.533) simultaneously, we obtain the following relationships:

(1) the relationship between solitary wavenumber and solitary frequency:

$$\omega = -k^3, \tag{5.534}$$

(2) the relationship between solitary amplitude and solitary wavenumber:

$$A = k^2B, \tag{5.535}$$

(3) the morphofactor:

$$B^2 = 4, \tag{5.536}$$

so the solitary wave solution can be readily obtained, which reads

$$u(x,t) = \frac{2k^2}{\exp(kx - k^3t + D) + \exp[-(kx - k^3t + D)] + 2}, \tag{5.537}$$

or

$$u(x,t) = \frac{-2k^2}{\exp(kx - k^3t + D) + \exp[-(kx - k^3t + D)] - 2}, \tag{5.538}$$

which are, by chance, exact solutions.

5.11.5. Ancient Chinese Inequality and Application

In an ancient history book, it is write (see [111]):

> *He Chengtian uses 26/49 as the strong, and 9/17 as the weak. Among the strong and the weak, Chengtian tries to find a more accurate denominator of the fractional day of the Moon. Chengtian obtains 752 as the denominator by using the 15 and 1 as weighting factors, respectively, for the strong and the weak. No other calendar can reach such a high accuracy after Chengtian, who uses heuristically the strong and weak weighting factors.*

The statement is rather cryptic, and in modern mathematical term, the statement can be explained as follows.

According to the observation data, He Chengtian (369?~447AD) finds that

$$29\frac{26}{49}\ \text{days} > 1\ \text{Moon} > 29\frac{9}{17}\ \text{days}. \tag{5.539}$$

Using the weighting factors (15 and 1), He Chengtian obtains

$$\text{The fractional day} = \frac{26 \times 15 + 9 \times 1}{49 \times 15 + 17 \times 1} = \frac{399}{752}, \tag{5.540}$$

so

$$1\ \text{Moon} = 29\frac{399}{752}\ \text{days}. \tag{5.541}$$

He Chengtian actually uses the following inequality.

If

$$\frac{a}{b} < x < \frac{d}{c}, \tag{5.542}$$

where $a, b, c,$ and d are real numbers, then

$$\frac{a}{b} < \frac{ma + nd}{mb + nc} < \frac{d}{c}, \tag{5.543}$$

and x is approximated by

$$x = \frac{ma + nd}{mb + nc}, \tag{5.544}$$

where m and n are weighting factors.

The above inequity was developed by the max-min approach to nonlinear oscillators [112]. Hereby we use the inequity for a chemical problem.

A mass balance on a differential volume element of porous medium for a spherical catalyst pellet gives

$$\frac{\partial c'}{\partial t} = \nabla \cdot D_e \nabla c' - R_A, \tag{5.545}$$

where c' is the chemical reactant concentration, $-R_A$ is the rate of reaction per unit volume, and D_e is the effective diffusion coefficient for reactant.

Assuming the coupled process of diffusion and reaction is steady, we reduce (5.545) to the form

$$D_e\left(\frac{\partial^2 c'}{\partial r^2} + \frac{2}{r}\frac{\partial c'}{\partial r}\right) = R_A. \tag{5.546}$$

The boundary conditions are $\partial c'/\partial r = 0,\ r = 0$ (center of catalyst), and $c' = c_s,\ r = r_0$ (surface of catalyst).

Introducing dimensionless variables $C = c'/c_s$ and $R = r/r_0$, we have the following nonlinear equation for nth-order and irreversible reaction at isothermal condition:

$$\frac{\partial^2 C}{\partial R^2} + \frac{2}{R}\frac{\partial C}{\partial R} = \phi^2 C^n,\quad \frac{\partial C}{\partial R}(0) = 0,\ C(1) = 1, \tag{5.547}$$

where ϕ is the Thiele modulus, defined as $\phi = \sqrt{r_0^2 k_v c_s^{n-1}/D_e}$, where k_v is the reaction constant.

Equation (5.547) can also be written in Cartesian geometry, and it is easy to transform Cartesian geometry to cylindrical or spherical geometry.

To investigate the nature of the solution of (5.547), consider the analogous linear equation

$$\frac{\partial^2 C}{\partial R^2} + \frac{2}{R}\frac{\partial C}{\partial R} = k,\quad \frac{\partial C}{\partial R}(0) = 0,\ C(1) = 1. \tag{5.548}$$

Multiplying R^2 on both sides of (5.548), we have

$$\frac{\partial}{\partial R}\left(R^2\frac{\partial C}{\partial R}\right) = kR^2,\quad \frac{\partial C}{\partial R}(0) = 0,\ C(1) = 1. \tag{5.549}$$

Its solution is

$$C = 1 + \frac{1}{6}k\left(R^2 - 1\right). \tag{5.550}$$

Now we rewrite (5.544) in the form

$$\frac{\partial^2 C}{\partial R^2} + \frac{2}{R}\frac{\partial C}{\partial R} = g(C), \tag{5.551}$$

where $g(C) = \phi^2 C^n$.

The left side of (5.551) is analogous to k in (5.548).

Observe that $0 < C < 1$, so C is never less than the solution of the following initial value problem:

$$\frac{\partial^2 C}{\partial R^2} + \frac{2}{R}\frac{\partial C}{\partial R} = \phi^2 C_{\max}^n = \phi^2. \tag{5.552}$$

The solution of (5.552) reads

$$C = 1 + \frac{1}{6}\phi^2\left(R^2 - 1\right). \tag{5.553}$$

So it follows that (see Figure 4)

$$\frac{6 + \phi^2(R^2 - 1)}{6} < C < 1. \tag{5.554}$$

According to He Chengtian's interpolation, we have

$$C = \frac{6p + p\phi^2(R^2 - 1) + q}{6p + q}, \tag{5.555}$$

or

$$C = 1 + \xi\phi^2\left(R^2 - 1\right), \tag{5.556}$$

where p and q are weighting factors, $\xi = p/(6p + q)$. The free parameter, ξ, in (5.556) can be identified via various methods. Hereby we illustrate a simple approach by the residual method. Substituting (5.556) into (5.544), we obtain the following residual equation:

$$\Pi(R,\xi) = \frac{\partial^2 C}{\partial R^2} + \frac{2}{R}\frac{\partial C}{\partial R} - \phi^2 C^n = 2\xi\phi^2 + 4\xi\phi^2 - \phi^2\left[1 + \xi\phi^2\left(R^2 - 1\right)\right]^n. \tag{5.557}$$

Chemists and technologists always want to have an accurate halfway through the change. Assume that the above residual vanishes completely at $C = 1/2$, that is, $1 + \xi\phi^2(R^2 - 1) = 1/2$, then we have

$$\Pi(R,\xi)|_{R=R^*} = 6\xi\phi^2 - \frac{1}{2^n}\phi^2, \tag{5.558}$$

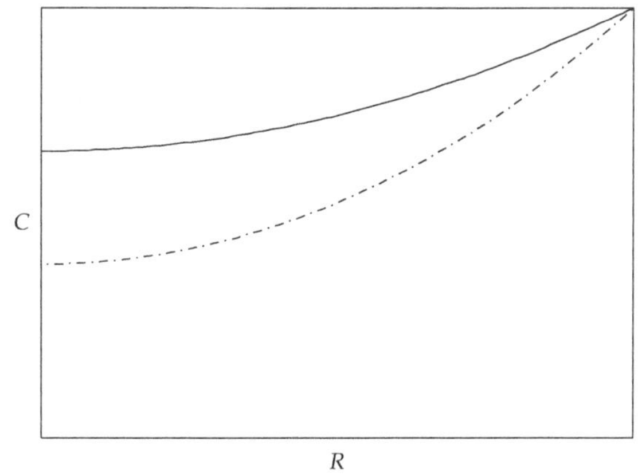

Figure 4: Qualitative sketch for the diffusion reaction in spherical porous catalyst. The dashed line presents an initial guess $C = 1 + \phi^2(R^2 - 1)/6$, and the continued line presents the real solution.

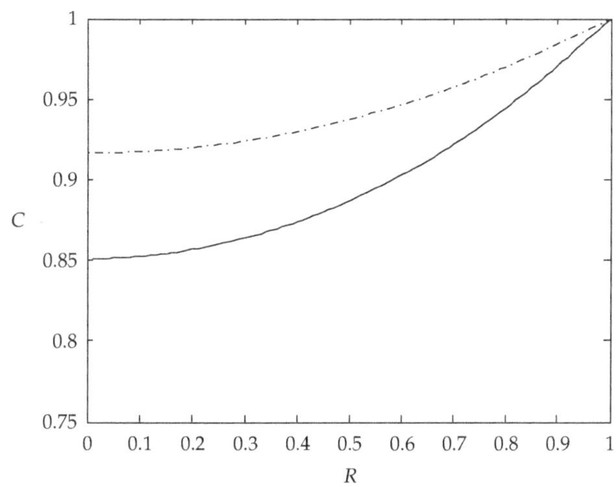

Figure 5: Comparison between exact solution and the approximate solution ($n = 1, \phi = 1$). The dashed line presents an approximate solution, and the continued line presents the exact solution. The maximal error is about 7.8%.

where $R = R^*$ is the root of the equation $1 + \xi\phi^2(R^2 - 1) = 1/2$. From (5.558), the unknown ξ can be identified as

$$\xi = \frac{2^{-n}}{6}. \tag{5.559}$$

Finally, we obtain the following design formulation:

$$C(R) = 1 + \frac{2^{-n}}{6}\phi^2\left(R^2 - 1\right). \tag{5.560}$$

Comparison with exact solution is illustrated in Figure 5.

In order to identify the weighting factors, p and q, in (5.555), we proceed the same manipulation as illustrated above. Substituting (5.555) into (5.544), we obtain a residual equation:

$$\Pi(R,p,q) = \frac{\partial^2 C}{\partial R^2} + \frac{2}{R}\frac{\partial C}{\partial R} - \phi^2 C^n = \frac{2p\phi^2}{6p+q} + \frac{4p\phi^2}{6p+q} - \phi^2\left(\frac{6p + p\phi^2(R^2-1)+q}{6p+q}\right)^n = 0. \tag{5.561}$$

Locating at $R = 0$ and $R = 1/2$, we obtain

$$\Pi(0,p,q) = \frac{2p\phi^2}{6p+q} + \frac{4p\phi^2}{6p+q} - \phi^2\left(\frac{6p - p\phi^2 + q}{6p+q}\right)^n = 0, \tag{5.562}$$

$$\Pi\left(\frac{1}{2},p,q\right) = \frac{2p\phi^2}{6p+q} + \frac{4p\phi^2}{6p+q} - \phi^2\left(\frac{6p - (3/4)p\phi^2 + q}{6p+q}\right)^n = 0. \tag{5.563}$$

Solving (5.562) and (5.563) simultaneously, the values of p and q can be readily determined.

The accuracy of vanishing residuals depends upon the trial functions, (5.555) and (5.556), which are derived from an analogous linear equation, so the obtained solutions are valid for the whole solution domain, keeping maximal accuracy at the location points.

The preceding analysis has the virtue of utter simplicity, and the illustrating example shows that the suggested method is very effective and convenient in solving nonlinear equations. The suggested method can be readily applied to the search for analytical solutions for various nonlinear equations, which should have the character of monotonic increase or monotonic decrease. Many nonlinear problems arising in chemical engineering can be described by, for example, exponential decay/rise functions, so the ancient Chinese method—the He Chengtian interpolation—might be a powerful mathematical tool in chemical engineering.

6. Solitons in Lattice Systems

6.1. Physical Understanding of Differential-Difference Equations

Governing equations for continuum media are well established by using the Gauss' divergence theorems:

$$\iiint_V \nabla\phi dV = \oiint_{\partial V} n\phi dS, \tag{6.1}$$

$$\iiint_V \nabla\cdot A dV = \oiint_{\partial V} n\cdot A dS, \tag{6.2}$$

$$\iiint_V \nabla\times A dV = \oiint_{\partial V} n\times A dS. \tag{6.3}$$

Mathematically, the above Gauss' divergence theorems are invalid for fractal media, such as porous media, weaves. So the governing equations for noncontinuum media cannot be expressed by differential equations in space but can be done by difference equations.

Considering the continuous time, such problems can be described by differential-difference equations:

$$\frac{\partial u_n}{\partial t} = F(u_{n-1}, u_n, u_{n+1}),$$
$$\frac{\partial u_n}{\partial t} = F\left(\frac{\partial u_{n-1}}{\partial x}, \frac{\partial u_n}{\partial x}, \frac{\partial u_{n-1}}{\partial x}, u_{n-1}, u_n, u_{n+1}\right), \tag{6.4}$$
$$\frac{\partial u_n}{\partial t} = F\left(\frac{\partial u_n}{\partial x}, \frac{\partial u_n}{\partial y}, \frac{\partial u_n}{\partial z}, u_{n-1}, u_n, u_{n+1}\right).$$

Unlike difference equations which are fully discretized, DDEs are semidiscretized with some of their spatial variables discretized, while time is usually kept continuous.

Solitons and compactons for difference-differential equations have caught much attention due to the fact that discrete spacetime may be the most radical and logical viewpoint of reality [113].

A better physical understanding of differential-difference equations can be obtained by considering the flow through a lattice where the conservation of mass requires [114]

$$\frac{d\rho_i}{dt} + \rho_{i+1}u_{i+1} - \rho_{i-1}u_{i-1} = 0, \tag{6.5}$$

with ρ_i and u_i being, respectively, the gas density and velocity at the ith lattice point (see Figure 6).

6.2. *Exp-Function Method*

Consider a hybrid-lattice system [115]

$$\frac{\partial u_n}{\partial t} = \left(1 + \alpha u_n + \beta u_n^2\right)(u_{n-1} - u_{n+1}), \tag{6.6}$$

where α and β are constants.

Let $u_n = u_n(\xi)$, $\xi = dn + c_1 x + c_2 t + \xi_0$, then (6.6) becomes

$$c_2 u_n' = \left(1 + \alpha u_n + \beta u_n^2\right)(u_{n-1} - u_{n+1}). \tag{6.7}$$

According to the exp-function method, we assume that

$$u_n(\xi) = \frac{a_1 \exp(\xi) + a_0 + a_{-1}\exp(-\xi)}{b_1 \exp(\xi) + b_0 + \exp(-\xi)},$$
$$u_{n-1}(\xi) = \frac{a_1 \exp(\xi - d) + a_0 + a_{-1}\exp(-\xi + d)}{b_1 \exp(\xi - d) + b_0 + \exp(-\xi + d)}, \tag{6.8}$$
$$u_{n+1}(\xi) = \frac{a_1 \exp(\xi + d) + a_0 + a_{-1}\exp(-\xi - d)}{b_1 \exp(\xi + d) + b_0 + \exp(-\xi - d)}.$$

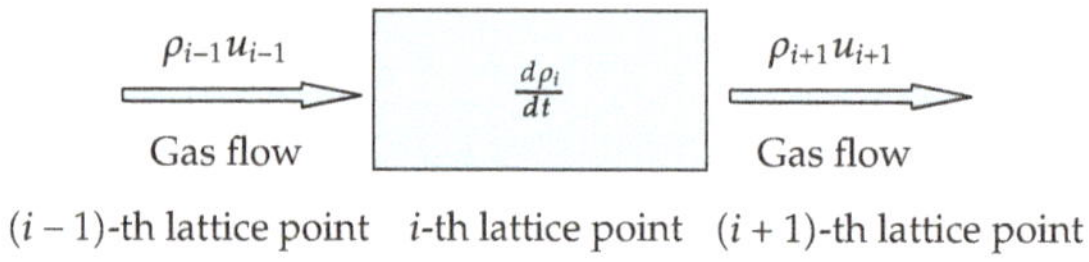

Figure 6: Conservation of mass.

Substituting (6.8) into (6.7), and by the help of Maple, clearing the denominator, and setting the coefficients of power terms like $\exp(j\xi), j = 1,2,\ldots$ to zero yield a system of algebraic equations, and we obtain some generalized exact solutions, one of which is expressed in the form [115]

$$u_n = \frac{\left((2\alpha\beta a_0^2 e^{2d})/(\alpha^2-4\beta)\left(e^{2d}-1\right)^2\right)\exp(\xi) + a_0 - (\alpha/2\beta)\exp(-\xi)}{-\left((4\alpha\beta^2 a_0^2 e^{2d})/(\alpha^2-4\beta)\left(e^{2d}-1\right)^2\right)\exp(\xi) + \exp(-\xi)}, \tag{6.9}$$

where $c_2 = (\alpha^2-4\beta)(e^{2d}-1)/(4\beta e^d)$, $\xi = dn + c_1 x + c_2 t + \xi_0$, and c_1 and a_0 are free parameters.

The application of the exp-function method to the discrete mKdV lattice and the discrete Schrödinger equation was discussed in [116, 117].

6.3. Variational Iteration Method

Consider the Volterra lattice equation [118]

$$\frac{\partial u_n}{\partial t} = u_n(u_{n+1} - u_{n-1}), \qquad u_n(0) = n. \tag{6.10}$$

Using the variational iteration method, an iteration formulation can be constructed as follows:

$$u_{n,m} = u_{n,m-1} - \int_0^t u_{n,m-1}\{u_{n+1,m-1} - u_{n-1,m-1}\}dt. \tag{6.11}$$

Begining with $(u_n)_0 = n$, we can obtain the following series solution [118]:

$$u_n(t) = n\left\{1 + 2t + 4t^2 + 8t^3 + 16t^4 + 32t^5 + 64t^6 + O\left(t^7\right)\right\}, \tag{6.12}$$

which converges to the exact solution $u_n(t) = n/(1-2t)$.

6.4. Homotopy Perturbation Method

Consider the discretized mKdV lattice equation [119]

$$\frac{du_n}{dt} = {u_n}^2(u_{n+1} - u_{n-1}), \qquad u_n(0) = 1 - \frac{1}{n^2}. \tag{6.13}$$

Construct a homotopy

$$(1-p)\left(\frac{du_n(t)}{dt}-\frac{du_{n,0}(t)}{dt}\right)+p\left(\frac{du_n(t)}{dt}-u_n^2(t)\cdot(u_{n+1}(t)-u_{n-1}(t))\right)=0. \quad (6.14)$$

Suppose the solution of (6.13) to be in the following form:

$$u_n(t)=u_{n,0}(t)+pu_{n,1}(t)+p^2u_{n,2}+\cdots. \quad (6.15)$$

By a simple calculation, beginning with $u_{n,0}(t)=1-1/n^2$, we have [119]

$$u_n(t)=1-\frac{1}{n^2}+\frac{4t}{n^3}-\frac{12t^2}{n^4}+\frac{32t^3}{n^5}-\frac{80t^4}{n^6}+\cdots, \quad (6.16)$$

which converges to the exact solution $u_n(t)=1-1/(n+2t)^2$.

6.5. Parameter-Expansion Method

We rewrite (6.13) in the form

$$\frac{du_n}{dt}=1\cdot u_n^2(u_{n+1}-u_{n-1}). \quad (6.17)$$

We expand, respectively, the solution and the coefficient, 1, in the forms

$$u_n(t)=u_{n,0}(t)+pu_{n,1}(t)+p^2u_{n,2}+\cdots,\quad p=1,$$
$$1=pa_1+p^2a_2+\cdots. \quad (6.18)$$

The solution procedure is given in Section 5.8.

6.6. Parameterized Perturbation Method

Parameterized perturbation method was proposed in 1999 [120]; it is useful to obtain exact and approximate solutions of nonlinear differential equations. The method requires no linearization or discretization; large computational work and round-off errors are avoided. It has been used to solve effectively, easily, and accurately a large class of nonlinear problems with approximations. These approximations converge rapidly to accurate solutions [120, 121].

We consider the following general nonlinear DDEs given in the form [122]:

$$\frac{du_n(t)}{dt}=f(u_{n-1},u_n,u_{n+1}), \quad (6.19)$$

with the initial condition $u_n(0)=A(n)$, where a, b, and c are constants, and f is a nonlinear function.

According to the PPM, an expanding parameter is introduced by a linear transformation

$$u_n = \varepsilon v_n, \tag{6.20}$$

where ε is the introduced perturbation parameter.

Substituting (6.20) into (6.19) results in

$$\begin{gathered} \frac{dv_n(t)}{dt} = f(\varepsilon; v_{n-1}, v_n, v_{n+1}), \\ v_n(0) = \frac{A(n)}{\varepsilon}, \end{gathered} \tag{6.21}$$

supposing that the solution of (6.21) can be expressed in the form

$$v_n = v_{n,0} + \varepsilon v_{n,1} + \varepsilon^2 v_{n,2} + \ldots, \tag{6.22}$$

and processing in a traditional fashion of perturbation technique.

In the following examples, we will illustrate the usefulness and effectiveness of the proposed technique.

Consider the following Volterra equation:

$$\frac{du_n}{dt} = u_n(u_{n+1} - u_{n-1}), \tag{6.23}$$

with the initial condition

$$u_n(0) = n. \tag{6.24}$$

Using the transformation (6.20), the original equation (6.23) becomes

$$\begin{gathered} \frac{dv_n}{dt} = \varepsilon v_n(v_{n+1} - v_{n-1}), \\ v_n(0) = \frac{n}{\varepsilon}. \end{gathered} \tag{6.25}$$

Substituting (6.22) into (6.25) and equating coefficients of like powers of ε yield the following equations:

$$\begin{gathered} \frac{dv_{n,0}}{dt} = 0, \qquad v_{n,0}(0) = v_n(0) = \frac{n}{\varepsilon}, \\ \frac{dv_{n,i}}{dt} = \sum_{k=0}^{i-1} v_{n,k}(v_{n+1,i-k-1} - v_{n-1,i-k-1}), \quad v_{n,i}(0) = 0,\ i = 1, 2, \ldots. \end{gathered} \tag{6.26}$$

We can start with $v_{n,0}(0) = n/\varepsilon$, and we obtain the following successive approximations:

$$v_{n,k}(t) = \frac{2^k n}{\varepsilon^{k+1}} t^k, \quad k = 0, 1, 2, \ldots. \tag{6.27}$$

Hence, the solution series in general gives

$$v_n = \frac{n}{\varepsilon}\left(1 + 2t + 4t^2 + 8t^3 + 16t^4 + \cdots\right), \tag{6.28}$$

then

$$u_n = \varepsilon v_n = n\left(1 + 2t + 4t^2 + 8t^3 + 16t^4 + \cdots\right). \tag{6.29}$$

The closed form of the series is

$$u_n(t) = \frac{n}{1 - 2t}, \tag{6.30}$$

which gives the exact solution of the problem.

6.7. Solitary-Solution Formulation [123]

Suppose that the differential-difference equation we discuss in this paper is in the following nonlinear polynomial form:

$$\frac{du_n(t)}{dt} = f(u_{n-1}, u_n, u_{n+1}), \tag{6.31}$$

where $u_n = u(n, t)$ is a dependent variable; t is a continuous variable; $n, p_i \in Z$.

Using the basic idea of the ancient Chinese algorithm, we choose two trial functions in the forms

$$\begin{aligned} u_{n,1}(n,t) &= f(\xi_n + \omega_1 t), \\ u_{n,2}(n,t) &= g(\xi_n + \omega_2 t), \end{aligned} \tag{6.32}$$

where $\xi_n = nd + \xi_0$, ξ_0 is arbitrary, and f and g are known functions. If a periodic solution is searched for, f and g must be periodic functions; if a solitary solution is solved, f and g must be of solitary structures. In this paper, a bell solitary solution of a differential-difference equation is considered, and trial functions are chosen as follows:

$$\begin{aligned} u_{n,1}(n,t) &= \frac{A}{e^{\xi_n + \omega_1 t} + e^{-(\xi_n + \omega_1 t)} + B}, \\ u_{n,2}(n,t) &= \frac{A}{e^{\xi_n + \omega t} + e^{-(\xi_n + \omega t)} + B}. \end{aligned} \tag{6.33}$$

For u_n, u_{n-1} and u_{n+1} should be compatible, then

$$u_{n-1,1}(n,t) = \frac{A}{e^{\xi_n-d+t} + e^{-(\xi_n-d+t)} + B}, \tag{6.34}$$

$$u_{n+1,1}(n,t) = \frac{A}{e^{\xi_n+d+t} + e^{-(\xi_n+d+t)} + B}, \tag{6.35}$$

$$u_{n-1,2}(n,t) = \frac{A}{e^{\xi_n-d+\omega t} + e^{-(\xi_n-d+\omega t)} + B}, \tag{6.36}$$

$$u_{n+1,2}(n,t) = \frac{A}{e^{\xi_n+d+\omega t} + e^{-(\xi_n+d+\omega t)} + B}. \tag{6.37}$$

6.8. Solution Procedure

Step 1. Define the residual function

$$\widetilde{R}(t) = \frac{du_n(t)}{dt} - f(u_{n-1}, u_n, u_{n+1}). \tag{6.38}$$

Substituting (6.40)~(6.43) into (6.37), we can obtain, respectively, the residual functions $\widetilde{R}_1$ and $\widetilde{R}_2$.

Step 2. Substitute $\widetilde{R}_1$ and $\widetilde{R}_2$ into the following equation:

$$\omega^2 = \frac{\omega_1^2 \widetilde{R}_2(0) - \omega_2^2 \widetilde{R}_1(0)}{\widetilde{R}_2(0) - \widetilde{R}_1(0)}, \tag{6.39}$$

where $\omega_2 = \omega$.

Step 3. Combining the coefficients of e^{ξ_n} in (6.45), and setting them to be zero, we can solve the algebraic equations to find the values of ω, A, and B. Finally, an explicit solution is obtained.

The famous mKdV lattice equation reads

$$\frac{du_n}{dt} = \left(\alpha - u_n^2\right)(u_{n-1} - u_{n+1}), \tag{6.40}$$

where $\alpha \neq 0$.

Substituting (6.40)~(6.43) into (6.46), we obtain, respectively, the following residual functions $\tilde{R}_1$ and $\tilde{R}_2$, and using the solitary-solution formulation, (6.45), we have

$$\begin{aligned}
&e^{3\xi_n}\left(\omega^3+\omega^2\alpha e^{-d}-\omega^2\alpha e^{d}-\omega-\alpha e^{-d}+\alpha e^{d}\right)-e^{-3\xi_n}\left(\omega^3+\omega^2\alpha e^{-d}-\omega^2\alpha e^{d}-\omega-\alpha e^{-d}+\alpha e^{d}\right)\\
&\quad+e^{2\xi_n}B\left(\omega^3e^{-d}+\omega^3e^{d}-2\omega^2\alpha e^{d}+2\omega^2\alpha e^{-d}-\omega e^{d}-\omega e^{-d}+2\alpha e^{d}-2\alpha e^{-d}\right)\\
&\quad-e^{-2\xi_n}B\left(\omega^3e^{-d}+\omega^3e^{d}-2\omega^2\alpha e^{d}+2\omega^2\alpha e^{-d}-\omega e^{d}-\omega e^{-d}+2\alpha e^{d}-2\alpha e^{-d}\right)\\
&\quad+e^{\xi_n}\Big(\omega^3B^2-\omega^3+\omega^3e^{-2d}+\omega^3e^{2d}+\omega^2A^2e^{d}-\omega^2A^2e^{-d}+\omega^2\alpha e^{-d}-\omega^2\alpha e^{d}+\omega^2\alpha B^2e^{-d}\\
&\qquad-\omega^2\alpha B^2e^{d}-\omega B^2+\omega-\omega e^{2d}-\omega e^{-2d}+A^2e^{-d}-A^2e^{d}+\alpha B^2e^{d}-\alpha B^2e^{-d}+\alpha e^{d}-\alpha e^{-d}\Big)\\
&\quad-e^{-\xi_n}\Big(\omega^3B^2-\omega^3+\omega^3e^{-2d}+\omega^3e^{2d}+\omega^2A^2e^{d}-\omega^2A^2e^{-d}+\omega^2\alpha e^{-d}-\omega^2\alpha e^{d}+\omega^2\alpha B^2e^{-d}\\
&\qquad-\omega^2\alpha B^2e^{d}-\omega B^2+\omega-\omega e^{2d}-\omega e^{-2d}+A^2e^{-d}-A^2e^{d}+\alpha B^2e^{d}-\alpha B^2e^{-d}+\alpha e^{d}-\alpha e^{-d}\Big)=0.
\end{aligned}\tag{6.41}$$

Setting the coefficients of $e^{i\xi_n}$ to be zero, we have

$$\begin{aligned}
&\omega^3+\omega^2\alpha e^{-d}-\omega^2\alpha e^{d}-\omega-\alpha e^{-d}+\alpha e^{d}=0,\\
&B\left(\omega^3e^{-d}+\omega^3e^{d}-2\omega^2\alpha e^{d}+2\omega^2\alpha e^{-d}-\omega e^{d}-\omega e^{-d}+2\alpha e^{d}-2\alpha e^{-d}\right)=0,\\
&\omega^3B^2-\omega^3+\omega^3e^{-2d}+\omega^3e^{2d}+\omega^2A^2e^{d}-\omega^2A^2e^{-d}+\omega^2\alpha e^{-d}-\omega^2\alpha e^{d}+\omega^2\alpha B^2e^{-d}\\
&\quad-\omega^2\alpha B^2e^{d}-\omega B^2+\omega-\omega e^{2d}-\omega e^{-2d}+A^2e^{-d}-A^2e^{d}+\alpha B^2e^{d}-\alpha B^2e^{-d}+\alpha e^{d}-\alpha e^{-d}=0.
\end{aligned}\tag{6.42}$$

Solving (6.48)–(6.49) simultaneously, we have

$$\omega=2\alpha\sinh(d),\tag{6.43}$$

$$A=2\sqrt{-\alpha}\sinh(d),\tag{6.44}$$

$$B=0.\tag{6.45}$$

We, therefore, obtain the following needed solitary solution:

$$\begin{aligned}
u_n&=\frac{2\sqrt{-\alpha}\sinh(d)}{e^{\xi_n+2\alpha\sinh(d)t}+e^{-(\xi_n+2\alpha\sinh(d)t)}}\\
&=\sqrt{-\alpha}\sinh(d)\operatorname{sech}\{2\alpha\sinh(d)t+nd+\xi_0\}.
\end{aligned}\tag{6.46}$$

We can also use the following assumptions for solitary solutions:

$$u_{n,1}(n,t) = \frac{d}{ae^{\xi_n+\omega_1 t} + be^{-(\xi_n+\omega_1 t)} + c}, \tag{6.47}$$

$$u_{n,2}(n,t) = \frac{\sum_{n=-c}^{d} a_n \exp(\xi_n + \omega t)}{\sum_{m=-f}^{g} b_m \exp(\xi_n + \omega t)}. \tag{6.48}$$

For compact-like solutions, we can assume that the solution has the form

$$u_{n,1}(n,t) = \frac{d}{a + \cos^2(\xi_n + \omega_1 t)}, \tag{6.49}$$

$$u_0(x,t) = \frac{a\sin^2(kx + wt)}{b + c\sin^2(kx + wt)}. \tag{6.50}$$

7. Fractional Differential Equations

7.1. Physical Understanding of Fractional Differential Equations

Fractional differential equations have caught much attention recently due to the exact description of nonlinear phenomena. No analytical method was available before 1998 for such equations even for linear fractional differential equations.

In 1998, the variational iteration method was first proposed to solve fractional differential equations with greatest success, see [124]. Following the above idea, Draganescu, Momani, and Odibat applied the variational iteration method to more complex fractional differential equations, showing effectiveness and accuracy of the used method, see [125, 126]. In 2002, the Adomian method was suggested to solve fractional differential equations [127]. But many researchers found that it is very difficult to calculate the Adomian polynomial, see comments in [128]. Ghorbani and Saberi-Nadjafi suggested a very simple method for calculating the Adomian polynomial using the homotopy perturbation method [129], and He polynomial should be used instead of Adomian polynomial [130].

In 2007, Momani and Odibat [131] applied the homotopy perturbation method to fractional differential equations and revealed that the homotopy perturbation method is an alternative analytical method for fractional differential equations.

Although the fractional calculus was invented by Newton and Leibnitz over three centuries ago, it only became a hot topic recently owing to the development of the computer and its exact description of many real-life problems. To give a physical interpretation of the fractional calculus, we begin with a simple function

$$y = x^2. \tag{7.1}$$

When $x = x_1$ and $x = x_2$, we have

$$y_1 = x_1^2, \qquad y_2 = x_2^2. \tag{7.2}$$

We therefore have the difference

$$\Delta y = y_1 - y_2 = x_1^2 - x_2^2 = (x_1 - x_2)(x_1 + x_2) = (x_1 + x_2)\Delta x. \tag{7.3}$$

In case $x_1 \to x_2$ or $\Delta x \to 0$, we have the differential

$$dy = 2xdx. \tag{7.4}$$

The above derivation, however, is only valid for continuous functions. To show its applications, we consider the action in String theory [132]

$$S = mc\int ds, \tag{7.5}$$

where m is the mass of a particle, c is the speed of light, and ds is the relativistic metric that can be expressed in the form

$$ds = \sqrt{c^2dt^2 - dx^2 - dy^2 - dz^2} = c\sqrt{1 - \frac{u^2}{c^2}}dt. \tag{7.6}$$

Substitution of (7.6) into (7.5) leads to

$$S = \int mc^2\sqrt{1 - \frac{u^2}{c^2}}dt, \tag{7.7}$$

from which the equation of free motion is obtained in the usual way

$$\frac{d}{dt}\left(\frac{mu}{\sqrt{1 - u^2/c^2}}\right) = 0. \tag{7.8}$$

The above derivation assumes that space and time are both continuous. The distance between two points, for example, cannot be expressed in the form of (7.6) in a discontinuous world (e.g., in the Julia set).

Now consider a plane with fractal structure (see Figure 7). The shortest path between two points is not a line, and we have

$$ds_E = kds^D, \tag{7.9}$$

where ds_E is the actual distance between two points (discontinuous line in Figure 7), ds is the line distance between two points (continuous line in Figure 7), D is the fractal dimension, and k is a constant.

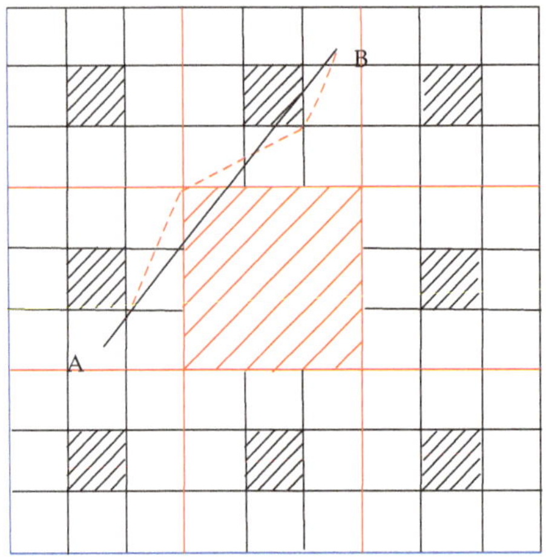

Figure 7: The distance between two points in a discontinuous spacetime.

The action in a discontinuous spacetime can therefore be written in the following form using fractional calculus:

$$S = mc\int ds_E = \int mc^{1+D}k\left(1 - \frac{u^2}{c^2}\right)^{D/2} dt^D. \tag{7.10}$$

Fractional calculus is therefore valid for discontinuous problems. Now we consider a well-known predator-prey model (the Lotka-Volterra equation) [133]

$$\begin{aligned} \frac{dx}{dt} &= x(a - by), \\ \frac{dy}{dt} &= -y(c - dy), \end{aligned} \tag{7.11}$$

where y is the number of predators (e.g., wolves), x is the number of its prey (e.g., rabbits), and $a, b, c,$ and d are parameters representing the interaction of the two species.

In general, the growth of the two populations is discontinuous, and a simple modification of the predator-prey model is to replace dy/dt and dx/dt by fractional derivatives

$$\begin{aligned} \frac{D^{\alpha}x}{Dt^{\alpha}} &= x(a - by), \\ \frac{D^{\beta}y}{Dt^{\beta}} &= -y(c - dy), \end{aligned} \tag{7.12}$$

where the populations of the predator and prey may be greatly affected by the fractional orders, α and β.

7.2. Variational Iteration Method

Consider the following general fractional differential equation:

$$\frac{D^\alpha u}{Dt^\alpha} + f = 0. \tag{7.13a}$$

In the case $0 < \alpha < 1$, we rewrite (7.13a) in the form

$$\frac{du}{dt} + \frac{D^\alpha u}{Dt^\alpha} - \frac{du}{dt} + f = 0, \tag{7.13b}$$

and the variational iteration algorithms are given as follows:

$$\begin{aligned}
u_{n+1}(t) &= u_n(t) - \int_0^t \left(\frac{D^\alpha u_n}{Dt^\alpha} + f_n\right) ds, \\
u_{n+1}(t) &= u_0(t) - \int_0^t \left(\frac{D^\alpha u_n}{Dt^\alpha} - \frac{du_n}{dt} + f_n\right) ds, \\
u_{n+1}(t) &= u_0(t) - \int_0^t \left\{\left(\frac{D^\alpha u_n}{Dt^\alpha} - \frac{du_n}{dt} + f_n\right) - \left(\frac{D^\alpha u_{n-1}}{Dt^\alpha} - \frac{du_{n-1}}{dt} + f_{n-1}\right)\right\} ds.
\end{aligned} \tag{7.14}$$

In the case $1 < \alpha < 2$, the above iteration formulas are also valid. We can rewrite (7.13a) in the form

$$\frac{d^2u}{dt^2} + \frac{D^\alpha u}{Dt^\alpha} - \frac{d^2u}{dt^2} + f = 0, \tag{7.15}$$

and the following iteration formulae are suggested:

$$\begin{aligned}
u_{n+1}(t) &= u_n(t) + \int_0^t (s-t)\left(\frac{D^\alpha u_n}{Dt^\alpha} + f_n\right) ds, \\
u_{n+1}(t) &= u_0(t) + \int_0^t (s-t)\left(\frac{D^\alpha u_n}{Dt^\alpha} - \frac{d^2u_n}{dt^2} + f_n\right) ds, \\
u_{n+1}(t) &= u_n(t) + \int_0^t (s-t)\left\{\left(\frac{D^\alpha u_n}{Dt^\alpha} - \frac{d^2u_n}{dt^2} + f_n\right) - \left(\frac{D^\alpha u_{n-1}}{Dt^\alpha} - \frac{d^2u_{n-1}}{dt^2} + f_{n-1}\right)\right\} ds.
\end{aligned} \tag{7.16}$$

When α is close to 1, (7.14) is better, while (7.16) is recommended for α approaching 2.

For the case $N < \alpha < N+1$, where N is a natural number, the iteration formulas

$$u_{n+1}(t) = u_n(t) + (-1)^N \int_0^t \frac{1}{(N-1)!}(s-t)^{N-1}\left(\frac{D^\alpha u_n}{Dt^\alpha} + f_n\right)ds,$$

$$u_{n+1}(t) = u_0(t) + (-1)^N \int_0^t \frac{1}{(N-1)!}(s-t)^{N-1}\left(\frac{D^\alpha u_n}{Dt^\alpha} - \frac{d^N u_n}{dt^N} + f_n\right)ds,$$

$$\begin{aligned} u_{n+1}(t) &= u_n(t) + (-1)^N \\ &\times \int_0^t \frac{1}{(N-1)!}(s-t)^{N-1} \\ &\times \left\{\left(\frac{D^\alpha u_n}{Dt^\alpha} - \frac{d^N u_n}{dt^N} + f_n\right) - \left(\frac{D^\alpha u_{n-1}}{Dt^\alpha} - \frac{d^N u_{n-1}}{dt^N} + f_{n-1}\right)\right\}ds, \end{aligned} \tag{7.17}$$

or

$$u_{n+1}(t) = u_n(t) + (-1)^{N+1} \int_0^t \frac{1}{N!}(s-t)^{N}\left(\frac{D^\alpha u_n}{Dt^\alpha} + f_n\right)ds,$$

$$u_{n+1}(t) = u_0(t) + (-1)^{N+1} \int_0^t \frac{1}{N!}(s-t)^{N}\left(\frac{D^\alpha u_n}{Dt^\alpha} - \frac{d^{N+1} u_n}{dt^{N+1}} + f_n\right)ds,$$

$$\begin{aligned} u_{n+1}(t) &= u_n(t) + (-1)^{N+1} \\ &\times \int_0^t \frac{1}{N!}(s-t)^{N}\left\{\left(\frac{D^\alpha u_n}{Dt^\alpha} - \frac{d^{N+1} u_n}{dt^{N+1}} + f_n\right) - \left(\frac{D^\alpha u_{n-1}}{Dt^\alpha} - \frac{d^{N+1} u_{n-1}}{dt^{N+1}} + f_{n-1}\right)\right\}ds \end{aligned} \tag{7.18}$$

can be used. When α is close to N, (7.17) is recommended, and (7.18) works more effectively for α approaching $N+1$.

We consider the fractional calculus version of the standard vibration equation in one dimension as [133]

$$\frac{\partial^2 u}{\partial r^2} + \frac{1}{r}\frac{\partial u}{\partial r} = \frac{1}{c^2}\frac{\partial^\beta u}{\partial t^\beta}, \quad r \geq 0,\ t \geq 0,\ 1 < \beta \leq 2, \tag{7.19}$$

with initial conditions

$$u(r,0) = r^2, \qquad \frac{\partial}{\partial t}u(r,0) = cr. \tag{7.20}$$

Equation (7.19) can be written as

$$\frac{\partial^2 u}{\partial t^2} = c^2 \frac{\partial^{2-\beta}}{\partial t^{2-\beta}}\left[\frac{\partial^2 u}{\partial r^2} + \frac{1}{r}\frac{\partial u}{\partial r}\right]. \tag{7.21}$$

According to the variational iteration method, we consider the correction functional in t-direction in the following form:

$$u_{n+1}(r,t) = u_n(r,t) + \int_0^t (\xi - t)\left\{\frac{\partial^2 u_n(r,\xi)}{\partial \xi^2} - c^2 \frac{\partial^{2-\beta}}{\partial t^{2-\beta}}\left(\frac{\partial^2 \tilde{u}_n(r,\xi)}{\partial r^2} + \frac{1}{r}\frac{\partial \tilde{u}_n(r,\xi)}{\partial r}\right)\right\} d\xi. \quad (7.22)$$

Beginning with an initial approximation

$$u_0(r,t) = u(r,0) + t u_t(r,0) = r^2 + ctr, \quad (7.23)$$

we obtain [133]

$$\begin{aligned} u(r,t) &= \lim_{n \to \infty} u_n(r,t) \\ &= r^2 + ctr + \frac{4c^2 t^\beta}{\Gamma(\beta+1)} + \frac{c^3 t^{\beta+1}}{r\Gamma(\beta+2)} + \frac{c^5 t^{2\beta+1}}{r^3\Gamma(2\beta+2)} + \frac{9c^7 t^{3\beta+1}}{r^5\Gamma(3\beta+2)} + \cdots \\ &= r^2 + \frac{4c^2 t^\beta}{\Gamma(\beta+1)} + ctrE_{\beta,2}\left(\frac{c^2}{r^2} k t^\beta\right), \end{aligned} \quad (7.24)$$

where $k^n = [1 \times 3 \times 5 \times \ldots \times (2n-3)]^2$ and $E_{\beta,b}(t) = \sum_{n=0}^{\infty} t^n/\Gamma(n\beta + b)$ are the generalized Mittag-Leffler function.

Consider another nonlinear fractional differential equation [134]

$$D_*^\alpha u(x) + u'(x) + u^3(x) = 0, \quad x > 0, \quad (7.25)$$

where $1 < \alpha \leq 2$, subject to the initial condition

$$u(0) = 1, \qquad u'(0) = 0. \quad (7.26)$$

We can construct a correction functional in the form

$$u_{n+1}(x) = u_n(x) + \int_0^x \lambda(\xi)\left(D^{(2)} u_n(\xi) + M\tilde{u}_n(\xi) - D^{(2)}\tilde{u}_n(\xi) + D^\alpha \tilde{u}_n(\xi)\right) d\xi. \quad (7.27)$$

After identification of the Lagrange multiplier, we obtain the following iteration formulation:

$$u_{n+1}(x) = u_n(x) + \int_0^x (\xi - x)\left[D_*^\alpha u_n(\xi) + u_n'(\xi) + u^3(\xi)\right] d\xi. \quad (7.28)$$

Consequently, beginning with $u_0 = 1$, we find the following approximations:

$$
\begin{aligned}
u_0(x) &= 1,\\
u_1(x) &= 1 - \frac{x^2}{2},\\
u_2(x) &= 1 - x^2 + \frac{x^3}{6} + \frac{x^4}{8} - \frac{x^6}{40} + \frac{x^8}{448} - \frac{\Gamma(4-\alpha)\, x^{4-\alpha}}{\Gamma(3-\alpha)\Gamma(5-\alpha)} + \frac{x^{4-\alpha}}{\Gamma(4-\alpha)}.
\end{aligned}
\tag{7.29}
$$

7.3. Homotopy Perturbation Method

The homotopy perturbation method becomes an effective tool to fractional differential equations [134–137]. Consider the nonlinear time-fractional RLW equation [134]

$$
\frac{\partial^\alpha u}{\partial t^\alpha} - u_{xxt} + \left(\frac{u^2}{2}\right)_x = 0, \quad t > 0,
\tag{7.30}
$$

where $0 < \alpha \leq 1$, subject to the initial condition

$$
u(x,0) = x.
\tag{7.31}
$$

We can construct the following homotopy:

$$
\frac{\partial u}{\partial t} = p\left[\frac{\partial u}{\partial t} + u_{xxt} - \left(\frac{u^2}{2}\right)_x - D_t^\alpha u\right].
\tag{7.32}
$$

Processing the same solution procedure as the homotopy perturbation method, we obtain

$$
\begin{aligned}
u_0 &= x,\\
u_1 &= -xt,\\
u_2 &= x\left(-t + t^2 + \frac{t^{2-\alpha}}{\Gamma(3-\alpha)}\right),\\
u_3 &= x\left(-t + 2t^2 - t^3 + \frac{2t^{2-\alpha}}{\Gamma(3-\alpha)} - \frac{4t^{3-\alpha}}{\Gamma(4-\alpha)} - \frac{t^{3-2\alpha}}{\Gamma(4-2\alpha)}\right),
\end{aligned}
\tag{7.33}
$$

and the solution can be approximately expressed as

$$
u = u_0 + u_1 + u_2 + u_3 + \dots.
\tag{7.34}
$$

7.4. Fractional Complex Transform

The fractional complex transform [138, 139] was originally suggested to convert a fractional differential equation with Jumarie's modification of Riemann-Liouville derivative into its classical differential partner.

Consider the following general fractional differential equation:

$$\begin{gathered} f\left(u, u_t^{(\alpha)}, u_x^{(\beta)}, u_y^{(\gamma)}, u_z^{(\lambda)}, u_t^{(2\alpha)}, u_x^{(2\beta)}, u_y^{(2\gamma)}, u_z^{(2\lambda)}, \ldots\right) = 0, \\ 0 < \alpha \leq 1, \qquad 0 < \beta \leq 1, \qquad 0 < \gamma \leq 1, \qquad 0 < \lambda \leq 1, \end{gathered} \tag{7.35}$$

where $u_t^{(\alpha)} = D_t^\alpha u = D^\alpha u / Dt^\alpha$ denotes Jumarie's fractional derivation, which is a modified Riemann-Liouville derivative defined as [140]

$$D_t^\alpha u(t, x, y, z) = \frac{1}{\Gamma(1-\alpha)} \frac{d}{dt} \int_0^t (t-\xi)^{-\alpha} \left(u(\xi, x, y, z) - u(0, x, y, z)\right) d\xi, \tag{7.36}$$

where u is a continuous (but not necessarily differentiable) function

$$\frac{\partial^\alpha c}{\partial t^\alpha} = 0, \tag{7.37}$$

$$\frac{\partial^\alpha}{\partial t^\alpha}[cu] = c \frac{\partial^\alpha u}{\partial t^\alpha}, \tag{7.38}$$

$$\frac{\partial^\alpha t^\beta}{\partial t^\alpha} = \frac{\Gamma(1+\beta)}{\Gamma(1+\beta-\alpha)} t^{\beta-\alpha}, \quad \beta \geq \alpha > 0. \tag{7.39}$$

In our previous publications [138], there were some flaws. The solution procedure should be followed as follows.

Using the following transforms:

$$\begin{gathered} s = t^\alpha, \\ X = x^\beta, \\ Y = y^\gamma, \\ Z = z^\lambda, \end{gathered} \tag{7.40}$$

we have

$$\begin{aligned}
\frac{\partial^\alpha u}{\partial t^\alpha} &= \frac{\partial u}{\partial s}\cdot\frac{\partial^\beta s}{\partial x^\beta} = \sigma_s \frac{\partial u}{\partial s},\\
\frac{\partial^\beta u}{\partial x^\beta} &= \frac{\partial u}{\partial X}\cdot\frac{\partial^\beta X}{\partial x^\beta} = \sigma_X \frac{\partial u}{\partial X},\\
\frac{\partial^\gamma u}{\partial y^\gamma} &= \frac{\partial u}{\partial Y}\cdot\frac{\partial^\gamma Y}{\partial y^\gamma} = \sigma_Y \frac{\partial u}{\partial Y},\\
\frac{\partial^\lambda u}{\partial z^\lambda} &= \frac{\partial u}{\partial Z}\cdot\frac{\partial^\lambda Z}{\partial z^\lambda} = \sigma_Z \frac{\partial u}{\partial Z},
\end{aligned} \tag{7.41}$$

where σ_s, σ_X, σ_Y, and σ_Z are fractal indexes [139]. We can, therefore, easily convert fractional differential equations into partial differential equations, so that everyone familiar with advanced calculus can deal with fractional calculus without any difficulty.

To determine σ_s, we consider a special case $s = t^\alpha$ and $u = s^m$, and we have

$$\frac{\partial^\alpha u}{\partial t^\alpha} = \frac{\Gamma(1+m\alpha)\cdot t^{m\alpha-\alpha}}{\Gamma(1+m\alpha-\alpha)} = \sigma\frac{\partial u}{\partial s} = \sigma m t^{m\alpha-\alpha}. \tag{7.42}$$

We, therefore, can determine σ_s as follows:

$$\sigma_s = \frac{\Gamma(1+m\alpha)}{m\Gamma(1+m\alpha-\alpha)}. \tag{7.43}$$

Other fractal indexes ($\sigma_X, \sigma_Y, \sigma_Z$) can be determined in a similar way.

As an example, consider the fractional differential equation

$$\frac{\partial^\alpha u}{\partial t^\alpha} + Bu = 0, \quad 0 < \alpha < 1, \; u(0) = 1. \tag{7.44}$$

Assume that the solution can be expressed in a series in the form

$$u = \sum_{m=0}^{\infty} a_m s^m, \tag{7.45}$$

where $a_m (m = 0, 1, 2, 3, \ldots)$ are constants to be further determined.

Using the transform $s = t^\alpha$, and submitting (7.45) into (7.44), we have

$$\frac{\partial}{\partial s}\sum_{m=0}^{\infty} \sigma_{sm} a_m s^m + B\sum_{m=0}^{\infty} a_m s^m = 0, \tag{7.46}$$

or

$$\sum_{m=0}^{\infty} m\sigma_{sm} a_m s^{m-1} + B\sum_{m=0}^{\infty} a_m s^m = 0. \tag{7.47}$$

According to (7.43), the fractal index σ_{sm} can be determined as follows:

$$\sigma_{sm} = \frac{\Gamma(1+m\alpha)}{m\Gamma(1+m\alpha-\alpha)}. \tag{7.48}$$

From (7.47) and (7.48), we obtain

$$\frac{\Gamma(1+m\alpha)}{\Gamma(1+m\alpha-\alpha)}a_m + Ba_{m-1} = 0. \tag{7.49}$$

Generally, we begin with $u_0 = u(0) = 1$. After a simple calculation, we have

$$a_m = \frac{(-B)^m}{\Gamma(1+m\alpha)}. \tag{7.50}$$

We, therefore, obtain the following solution:

$$u(s) = \sum_{m=0}^{\infty} \frac{(-B)^m}{\Gamma(1+m\alpha)} s^m, \tag{7.51}$$

or

$$u(t) = \sum_{m=0}^{\infty} \frac{(-B)^m}{\Gamma(1+m\alpha)} t^{m\alpha} = E_\alpha(-Bt^\alpha), \tag{7.52}$$

where E_α is a Mittag-Leffler function defined as

$$E_\alpha(t) = \sum_{m=0}^{\infty} \frac{t^m}{\Gamma(1+m\alpha)}. \tag{7.53}$$

Equation (7.53) is the exact solution of the example.

7.5. Yang-Laplace Transform [141]

The local fractional functional analysis was first proposed by Yang [141]. The Yang-Laplace transform of $f(x)$ is defined as

$$L_\alpha\{f(x)\} = f_s^{L,\alpha}(s) := \frac{1}{\Gamma(1+\alpha)} \int_0^\infty E_\alpha(-s^\alpha x^\alpha) f(x) (dx)^\alpha, \quad 0 < \alpha \le 1, \tag{7.54}$$

and the inverse formula is

$$L_\alpha^{-1}\left(f_s^{L,\alpha}(s)\right) = f(t) =: \frac{1}{(2\pi)^\alpha} \int_{\beta-i\infty}^{\beta+i\infty} E_\alpha(s^\alpha x^\alpha) f_s^{L,\alpha}(s) (ds)^\alpha, \tag{7.55}$$

where $s^\alpha = \beta^\alpha + i^\alpha \infty^\alpha$ and $\mathrm{Re}(s^\alpha) = \beta^\alpha > 0^\alpha$.

Table of the Yang-Laplace Transform of Elementary Functions

$$
\begin{aligned}
&(1)\ L_\alpha\{\delta_\alpha(x)\} = 1.\\
&(2)\ L_\alpha\{\delta_\alpha(x-a)\} = E_\alpha(-a^\alpha s^\alpha).\\
&(3)\ L_\alpha\{H_\alpha(x)\} = \frac{1}{s^\alpha}.\\
&(4)\ L_\alpha\{H_\alpha(x)x^\alpha\} = \frac{1}{s^{2\alpha}}.\\
&(5)\ L_\alpha\left\{\delta_\alpha^{(n\alpha)}(x)\right\} = s^{n\alpha}.\\
&(6)\ L_\alpha\{1\} = \frac{1}{s^\alpha}.\\
&(7)\ L_\alpha\{E_\alpha(a^\alpha x^\alpha)\} = \frac{1}{s^\alpha - a^\alpha}.\\
&(8)\ L_\alpha\{x^\alpha E_\alpha(a^\alpha x^\alpha)\} = \frac{1}{(s-a)^{2\alpha}}.\\
&(9)\ L_\alpha\{\sin_\alpha(ax)^\alpha\} = \frac{a^\alpha}{s^{2\alpha} - a^{2\alpha}}.\\
&(10)\ L_\alpha\{\cos_\alpha(ax)^\alpha\} = \frac{s^\alpha}{s^{2\alpha} + a^{2\alpha}}.\\
&(11)\ L_\alpha\{E_\alpha(-b^\alpha x^\alpha)\sin_\alpha(ax)^\alpha\} = \frac{a^\alpha}{(s+b)^{2\alpha} + a^{2\alpha}}.\\
&(12)\ L_\alpha\{E_\alpha(-b^\alpha x^\alpha)\cos_\alpha(ax)^\alpha\} = \frac{a^\alpha + b^\alpha}{(s+b)^{2\alpha} + a^{2\alpha}}.\\
&(13)\ L_\alpha\left\{x^{k\alpha}\right\} = \frac{\Gamma(1+k\alpha)}{s^{(k+1)\alpha}}.\\
&(14)\ L_\alpha\left\{x^{k\alpha}E_\alpha(a^\alpha x^\alpha)\right\} = \frac{\Gamma(1+k\alpha)}{(s-a)^{(k+1)\alpha}}.\\
&(15)\ L_\alpha\{E_\alpha(a^\alpha x^\alpha) - E_\alpha(b^\alpha x^\alpha)\} = \frac{a^\alpha - b^\alpha}{(s-a)^\alpha (s-b)^\alpha}.
\end{aligned}
\tag{7.56}
$$

The Yang-Laplace transform is an effective mathematical tool for solving local fractional differential equations. Consider the following local fractional differential equation:

$$y^{(\alpha)}(t) + 2y(t) = E_\alpha(-t^\alpha), \quad 0 < \alpha \le 1, \tag{7.57}$$

with initial condition $y(t)|_{t=0} = 0$.

By the Yang-Laplace transform, (7.55) becomes

$$s^{\alpha} y_{s}^{L,\alpha}(s) + 2y_{s}^{L,\alpha}(s) = \frac{1}{1+s}. \tag{7.58}$$

Solving $y_{s}^{L,\alpha}$ from (7.58), we have

$$y_{s}^{L,\alpha}(\omega) = \frac{1}{1+s^{\alpha}} - \frac{1}{2+s^{\alpha}}. \tag{7.59}$$

The inverse Yang-Laplace transform of (7.59) gives

$$y(t) = E_{\alpha}(-t^{\alpha}) - E_{\alpha}(-2t^{\alpha}). \tag{7.60}$$

Consider another local fractional differential equation given by

$$\frac{\partial^{\alpha} u}{\partial y^{\alpha}} + k^{\alpha} u = 0, \quad y > 0,\ 0 < \alpha \leq 1,\ u(0) = 1. \tag{7.61}$$

Taking the Yang-Laplace transform, we have

$$s^{\alpha}\tilde{u}(s) + k^{\alpha}\tilde{u}(s) - u(0) = 0. \tag{7.62}$$

It is obvious that

$$\tilde{u}(s) = \frac{1}{s^{\alpha} + k^{\alpha}}. \tag{7.63}$$

The inverse Yang-Laplace transform results in the following solution:

$$u(y) = E_{\alpha}(k^{\alpha} y^{\alpha}). \tag{7.64}$$

7.6. Yang-Fourier Transform

Yang-Fourier transform is also an effective method for local fractional differential equations. The Yang-Fourier transform of $f(x)$ is given by

$$F_{\alpha}\{f(x)\} = f_{\omega}^{F,\alpha}(\omega) := \frac{1}{\Gamma(1+\alpha)} \int_{-\infty}^{\infty} E_{\alpha}(-i^{\alpha}\omega^{\alpha}x^{\alpha}) f(x)(dx)^{\alpha}, \tag{7.65}$$

and the inverse formula of the Yang-Fourier transform is

$$f(x) = F_{\alpha}^{-1}\left(f_{\omega}^{F,\alpha}(\omega)\right) := \frac{1}{(2\pi)^{\alpha}} \int_{-\infty}^{\infty} E_{\alpha}(i^{\alpha}\omega^{\alpha}x^{\alpha}) f_{\omega}^{F,\alpha}(\omega)(d\omega)^{\alpha}. \tag{7.66}$$

The basic properties of the Yang-Fourier transform are as follows:

$$F_\alpha\{af(x)+bg(x)\} = aF_\alpha\{f(x)\}+bF_\alpha\{g(x)\},$$
$$F_\alpha\left\{f^{(\alpha)}(x)\right\} = i^\alpha\omega^\alpha F_\alpha\{f(x)\}. \tag{7.67}$$

Table of Local Fractional Fourier Transforms

$$\begin{aligned}
&(1)\ F_\alpha\{\delta_\alpha(x)\} = 1.\\
&(2)\ F_\alpha\{1\} = \frac{(2\pi)^\alpha}{\Gamma(1+\alpha)}\delta_\alpha(\omega).\\
&(3)\ F_\alpha\{E_\alpha(i^\alpha\omega_0^\alpha x^\alpha)\} = \frac{(2\pi)^\alpha}{\Gamma(1+\alpha)}\delta_\alpha(\omega-\omega_0).\\
&(4)\ F_\alpha\{\delta_\alpha(x-x_0)f(x)\} = E_\alpha(-i^\alpha\omega^\alpha x_0^\alpha)f(x_0).\\
&(5)\ F_\alpha\{\delta_\alpha(x-x_0)\} = E_\alpha(-i^\alpha\omega^\alpha x_0^\alpha).\\
&(6)\ F_\alpha\{\cos_\alpha(ax)^\alpha\} = \pi^\alpha[\delta_\alpha(\omega+a)+\delta_\alpha(\omega-a)].\\
&(7)\ F_\alpha\{\sin_\alpha(ax)^\alpha\} = \pi^\alpha i^\alpha[\delta_\alpha(\omega+a)-\delta_\alpha(\omega-a)].\\
&(8)\ F_\alpha\{H_\alpha(x)E_\alpha(i^\alpha\omega_0^\alpha x^\alpha)\} = \frac{1}{i^\alpha(\omega-a)^\alpha}+\frac{(2\pi)^\alpha}{2\Gamma(1+\alpha)}\delta_\alpha(\omega-a).\\
&(9)\ F_\alpha\{|x|^\alpha\} = -\frac{2}{\omega^{2\alpha}}.\\
&(10)\ F_\alpha\left\{\delta_\alpha^{(n)}(x)\right\} = j^{n\alpha}\omega^{n\alpha}.\\
&(11)\ F_\alpha\{x^{n\alpha}\} = \frac{(2\pi)^\alpha}{\Gamma(1+\alpha)}\delta_\alpha^{(n\alpha)}(\omega).\\
&(12)\ F_\alpha\{H_\alpha(x)x^{n\alpha}\} = \frac{\Gamma(1+n\alpha)}{i^{n\alpha}\omega^{n\alpha}}+\frac{(2\pi)^\alpha}{2\Gamma(1+\alpha)}i^{n\alpha}\delta_\alpha^{(n\alpha)}(\omega).
\end{aligned} \tag{7.68}$$

Consider a local fractional differential equation

$$y^{(\alpha)}(t)+2y(t) = E_\alpha(-t^\alpha),\quad 0<\alpha\le 1, \tag{7.69}$$

with initial condition $y(0) = 0$.

Taking Yang-Fourier transform for (7.69),

$$i^\alpha\omega^\alpha y_\omega^{F,\alpha}(\omega)+2y_\omega^{F,\alpha}(\omega) = \frac{1}{1+i^\alpha\omega^\alpha}, \tag{7.70}$$

we have

$$y_{\omega}^{F,\alpha}(\omega) = \frac{1}{1 + i^{\alpha}\omega^{\alpha}} - \frac{1}{2 + i^{\alpha}\omega^{\alpha}}. \tag{7.71}$$

The inverse Yang-Fourier transform gives

$$f(t) = E_{\alpha}(-t^{\alpha}) - E_{\alpha}(-2t^{\alpha}). \tag{7.72}$$

Consider another local fractional differential equation

$$\frac{\partial^{\alpha} u}{\partial y^{\alpha}} + k^{\alpha} u = 0, \quad y \in \mathbb{R},\ 0 < \alpha \leq 1,\ u(0) = 1. \tag{7.73}$$

Taking the Yang-Fourier transform yields

$$i^{\alpha}\omega^{\alpha}\tilde{u}(\omega) + k^{\alpha}\tilde{u}(\omega) - 1 = 0, \tag{7.74}$$

or

$$\tilde{u}(\omega) = \frac{1}{i^{\alpha}\omega^{\alpha} + k^{\alpha}}. \tag{7.75}$$

The inverse Yang-Fourier transform gives the solution

$$u(y) = E_{\alpha}(k^{\alpha} y^{\alpha}). \tag{7.76}$$

7.7. Fractal Derivative and q-Derivative

The fractal derivative is defined by transforming the standard integer-dimensional space time (x, t) into a fractal space time [142]

$$\frac{du(t)}{dt^{D}} = \lim_{s \to t} \frac{u(t) - u(s)}{t^{D} - s^{D}}, \tag{7.77}$$

where D is the order of the fractal derivative. This definition is much simpler but lacks physical understanding.

As an example, consider the fractal derivative relaxation equation [142]

$$\frac{du}{dt^{D}} + Bu = 0, \quad 0 < D < 1,\ u(0) = 1, \tag{7.78}$$

whose analytical solution is [142]

$$u(t) = \exp\left(-Bt^{D}\right). \tag{7.79}$$

We write down a general fractal differential equation of the form

$$\frac{du}{dt^D} + f = 0, \tag{7.80}$$

and use the transformation $t^D = x$ to convert (7.80) into an ordinary differential equation

$$\frac{du}{dx} + f = 0, \tag{7.81}$$

so that the variational iteration algorithms above can be directly applied:

$$\begin{aligned} u_{n+1}(x) &= u_n(x) - \int_0^x \left(\frac{du_n}{dx} + f_n\right)dx, \\ u_{n+1}(x) &= u_0(x) - \int_0^x f_n dx, \\ u_{n+1}(x) &= u_0(x) - \int_0^x \{f_n - f_{n-1}\}dx, \end{aligned} \tag{7.82}$$

or

$$\begin{aligned} u_{n+1}\left(t^D\right) &= u_n\left(t^D\right) - \int_0^{t^D} \left(\frac{du_n}{dt^D} + f_n\right)Dt^{D-1}dt, \\ u_{n+1}\left(t^D\right) &= u_0\left(t^D\right) - \int_0^{t^D} f_n Dt^{D-1}dt, \\ u_{n+1}\left(t^D\right) &= u_0\left(t^D\right) - \int_0^{t^D} \{f_n - f_{n-1}\}Dt^{D-1}dx. \end{aligned} \tag{7.83}$$

As another example, consider the Nth-order fractal differential equation

$$\frac{d^N u}{dt^D} + f = 0, \tag{7.84}$$

which can be converted by the transformation

$$t^D = x^N \tag{7.85}$$

into the ordinary differential equation

$$\frac{d^N u}{dx^N} + f = 0, \tag{7.86}$$

and the variational iteration algorithms are

$$u_{n+1}(x) = u_n(x) + (-1)^N \int_0^x \frac{1}{(N-1)!}(s-x)^{N-1}\left(u_n^{(N)} + f_n\right)ds,$$
$$u_{n+1}(x) = u_0(x) + (-1)^N \int_0^t \frac{1}{(N-1)!}(s-x)^{N-1} f_n ds, \tag{7.87}$$
$$u_{n+1}(x) = u_n(x) + (-1)^N \int_0^t \frac{1}{(N-1)!}(s-t)^{N-1}(f_n - f_{n-1})ds.$$

The fractal derivative is simpler than its fractional counterpart in many applications, and it is also valid for discontinuous cases. We rewrite (7.77) in the form

$$\frac{du(x)}{dx^D} = \frac{1}{k}\lim_{A\to B}\frac{u(A)-u(B)}{\tilde{x}_A - \tilde{x}_B}, \tag{7.88}$$

where k is a constant, and A and B are arbitrary points in discontinuous space or spacetime (as shown in Figure 8). $(\tilde{x}_A, \tilde{x}_B)$ are called the fractal coordinates and are defined by

$$\tilde{x}_A = k(x_A - 0)^D = k(x_A)^D, \tag{7.89}$$

$$\tilde{x}_B = k(x_B - 0)^D = k(x_B)^D, \tag{7.90}$$

where (x_A, x_B) are the coordinates, and D is the fractional dimension in x-direction. Substituting (7.88) and (7.89) into (7.88), we obtain

$$\frac{du(x)}{dx^D} = \lim_{A\to B}\frac{u(A)-u(B)}{(x_A)^D - (x_B)^D}. \tag{7.91}$$

The fractal differential model is particularly suitable for describing discontinuous matter and is the preferred model for describing flow or heat conduction through porous media. For instance, the principle of mass conservation can be written in the form

$$\frac{\partial\rho}{\partial t} + k_1\frac{\partial(\rho u)}{\partial x^{D_1}} + k_2\frac{\partial(\rho v)}{\partial y^{D_2}} + k_3\frac{\partial(\rho w)}{\partial z^{D_3}} = 0, \tag{7.92}$$

where D_1, D_2, and D_3 are the fractal dimensions of porosity in the x, y, and z directions, respectively, and $k_i(i = 1,2,3)$ are constants that are related to the fractal dimensions. In particular, we have $k_i = 1$, when $D_i = 1$. Similarly, the momentum equation for one-dimensional porous flow can be written in the form

$$\frac{\partial u}{\partial t} + ku\frac{\partial u}{\partial x^D} = -\frac{k}{\rho}\frac{\partial P}{\partial x^D} + k\frac{\partial}{\partial x^D}\left(\mu k\frac{\partial u}{\partial x^D}\right), \tag{7.93}$$

when $D = 1$ and $k = 1$, and (7.93) turns out to be the classical one.

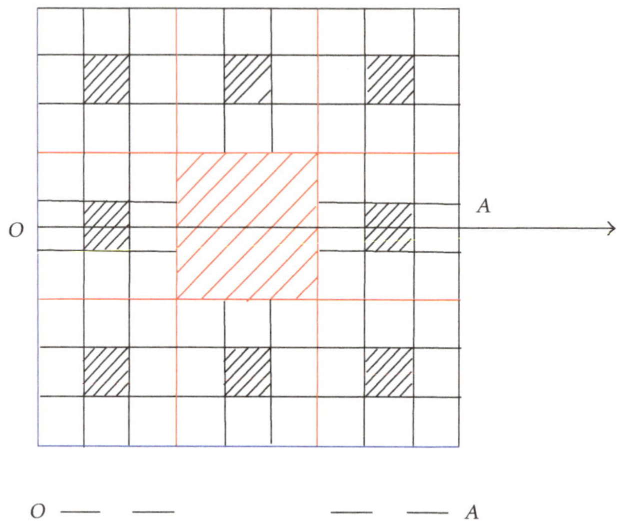

Figure 8: A schematic diagram of distance between O and A (i.e., the fractal coordinate of A) in a fractal space, and $D = \ln 2/\ln 3$, while the fractional dimensions for the plane are $\ln 8/\ln 3$.

The one-dimensional heat conduction equation in porous media can be expressed as

$$\frac{\partial T}{\partial t} + k\frac{\partial}{\partial x^D}\left(\mu k\frac{\partial T}{\partial x^D}\right) = 0, \tag{7.94}$$

where μ is the conduction coefficient and $k = 1$ when $D = 1$.

An oscillator swinging in a porous medium can also be described by fractal differential equations. For example, the Duffing equation with fractal damp can be expressed as

$$\frac{d^2u}{dt^2} + u + k\mu\frac{du}{dt^D} + \varepsilon u^3 = 0. \tag{7.95}$$

It is easy to establish fractal differential equations for discontinuous media by replacing $\partial/\partial x$ in the classical approach by $k\partial/\partial x^D$.

Now we consider a fractal media illustrated in Figure 9, and assume that the smallest measure is L_0, any discontinuity less than L_0 is ignored, then the distance between two points of A and B in Figure 1 can be expressed using fractal geometry. Hereby we introduce a new fractal derivative for engineering application:

$$\frac{Du(t)}{Dx^\alpha} = \lim_{\Delta x\to L_0}\frac{u(A)-u(B)}{\text{The distance between two points}} = \frac{du}{ds} = \lim_{\Delta x\to L_0}\frac{u(A)-u(B)}{kL_0^\alpha}, \tag{7.96}$$

where k is a constant, α is the fractal dimension, and the distance between two points in a discontinuous space can be expressed as $ds = kL_0^\alpha$.

Please note in the above definition that Δx does tend to zero, but to the smallest measure size, L_0.

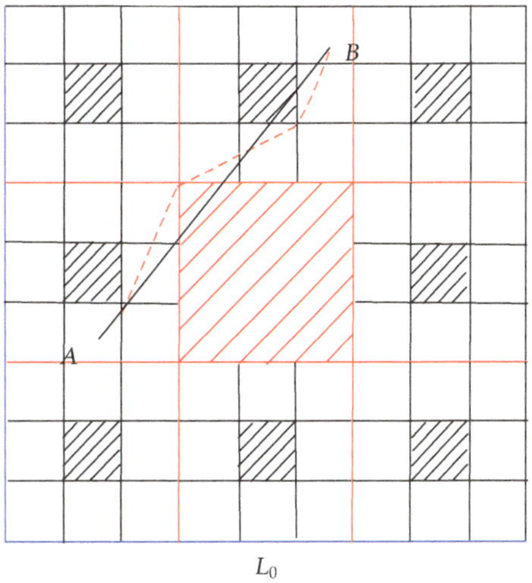

Figure 9: The distance between two points in a discontinuous spacetime.

Example 7.1. As a simple application, we consider the Fourier's law heat conduction, which reads

$$\frac{\partial q_h}{\partial t} = -c\frac{dT}{dn}, \tag{7.97}$$

where T is the thermal potential, and q_h is heat flow.

In the discontinuous media, the Fourier's law can be simply modified as

$$\frac{\partial q_h}{\partial t} = -c\frac{DT}{Dn^{\alpha}}, \tag{7.98}$$

where DT/Dn^{α} is a fractal derivative defined in (7.96).

The suggested fractal derivative is easy to be used for any discontinuous problems, and equations with fractal derivative can be easily solved using classical calculus.

Let $f(x;y;\ldots)$ be a multivariable real continuous function. The q-derivative and the partial q-derivative are defined by [143]:

$$\begin{aligned} D_q^x f(x) &= \frac{f(qx) - f(x)}{(q-1)x}, \\ \partial_q^x f(x;y;\ldots) &= \frac{f(qx;y;\ldots) - f(x;y;\ldots)}{(q-1)x}, \end{aligned} \tag{7.99}$$

and $\partial_q^x f(x;y;\ldots)|_{x=0} = \lim_{n\to\infty}(f(xq^n;y;\ldots) - f(0;y;\ldots))/xq^n$,

$$D_q^t u + f(u) = 0, \quad 0 < q < 1. \tag{7.100}$$

In the case $0 < \alpha < 1$, we rewrite (7.13a) in the form

$$\frac{du}{dt} + \left\{ D_q^t u + f(u) - \frac{du}{dt} \right\} = 0, \tag{7.101}$$

and the variational iteration algorithms are given as follows:

$$\begin{aligned} u_{n+1}(t) &= u_n(t) - \int_0^t \left(D_q^t u_n + f(u_n) \right) ds, \\ u_{n+1}(t) &= u_0(t) - \int_0^t \left(D_q^t u_n - \frac{du_n}{dt} + f(u_n) \right) ds, \\ u_{n+1}(t) &= u_0(t) - \int_0^t \left\{ \left(D_q^t u_n - \frac{du_n}{dt} + f(u_n) \right) - \left(\frac{D^\alpha u_{n-1}}{Dt^\alpha} - \frac{du_{n-1}}{dt} + f(u_{n-1}) \right) \right\} ds. \end{aligned} \tag{7.102}$$

In the case $1 < \alpha < 2$, the above iteration formulas are also valid. We can rewrite (7.100) in the form

$$\frac{d^2u}{dt^2} + \frac{D^\alpha u}{Dt^\alpha} - \frac{d^2u}{dt^2} + f = 0 \tag{7.103}$$

and the following iteration formulae are suggested

$$\begin{aligned} u_{n+1}(t) &= u_n(t) + \int_0^t (s-t) \left(\frac{D^\alpha u_n}{Dt^\alpha} + f_n \right) ds, \\ u_{n+1}(t) &= u_0(t) + \int_0^t (s-t) \left(\frac{D^\alpha u_n}{Dt^\alpha} - \frac{d^? u_n}{dt^2} + f_n \right) ds, \\ u_{n+1}(t) &= u_n(t) + \int_0^t (s-t) \left\{ \left(\frac{D^\alpha u_n}{Dt^\alpha} - \frac{d^2 u_n}{dt^2} + f_n \right) \right\} - \left(\frac{D^\alpha u_{n-1}}{Dt^\alpha} - \frac{d^2 u_{n-1}}{dt^2} + f_{n-1} \right) ds, \end{aligned} \tag{7.104}$$

when α is close to 1, (7.102) is better, while (7.104) is recommended for α approaching 2.

For the case $N < \alpha < N+1$, where N is a natural number, the iteration formulas are as follows

$$
\begin{aligned}
u_{n+1}(t) &= u_n(t) + (-1)^N \int_0^t \frac{1}{(N-1)!}(s-t)^{N-1}\left(\frac{D^\alpha u_n}{Dt^\alpha} + f_n\right)ds \\
u_{n+1}(t) &= u_0(t) + (-1)^N \int_0^t \frac{1}{(N-1)!}(s-t)^{N-1}\left(\frac{D^\alpha u_n}{Dt^\alpha} - \frac{d^N u_n}{dt^N} + f_n\right)ds \\
u_{n+1}(t) &= u_n(t) + (-1)^N + \int_0^t \frac{1}{(N-1)!}(s-t)^{N-1} \\
&\quad \times \left\{\left(\frac{D^\alpha u_n}{Dt^\alpha} - \frac{d^N u_n}{dt^N} + f_n\right) - \left(\frac{D^\alpha u_{n-1}}{Dt^\alpha} - \frac{d^N u_{n-1}}{dt^N} + f_{n-1}\right)\right\}ds,
\end{aligned}
\tag{7.105}
$$

or

$$
\begin{aligned}
u_{n+1}(t) &= u_n(t) + (-1)^{N+1} \int_0^t \frac{1}{N!}(s-t)^N\left(\frac{D^\alpha u_n}{Dt^\alpha} + f_n\right)ds, \\
u_{n+1}(t) &= u_0(t) + (-1)^{N+1} \int_0^t \frac{1}{N!}(s-t)^N\left(\frac{D^\alpha u_n}{Dt^\alpha} - \frac{d^{N+1} u_n}{dt^{N+1}} + f_n\right)ds, \\
u_{n+1}(t) &= u_n(t) + (-1)^{N+1} \\
&\quad \times \int_0^t \frac{1}{N!}(s-t)^N\left\{\left(\frac{D^\alpha u_n}{Dt^\alpha} - \frac{d^{N+1} u_n}{dt^{N+1}} + f_n\right) - \left(\frac{D^\alpha u_{n-1}}{Dt^\alpha} - \frac{d^{N+1} u_{n-1}}{dt^{N+1}} + f_{n-1}\right)\right\}ds.
\end{aligned}
\tag{7.106}
$$

8. Nonlinear Wave Transform for Nonlinear Wave Equations

The linear wave transform is widely used in nonlinear community to search for solitary solutions. Take the well-known KdV equation as an example

$$\frac{\partial u}{\partial t} - 6u\frac{\partial u}{\partial x} + \frac{\partial^3 u}{\partial x^3} = 0. \tag{8.1}$$

The general solution approach is to introduce a wave transform

$$\xi = x - Vt \tag{8.2}$$

to convert (8.1) into an ordinary differential equation. In the linear wave transform, V has a definite physical understanding, and it is the wave speed.

The nonlinear wave transform introduces a complex variable ξ defined as

$$\xi = a(t)x + b(t)y + c(t)z + d(t), \tag{8.3}$$

where $a, b, c,$ and d are unknown functions of time.

In order to elucidate the solution procedure, we consider the following equation:

$$u_t + t^2 u_x + stuu_x + rtu_{xxx} = 0, \tag{8.4}$$

where r and s are constants.

By a nonlinear wave transform,

$$\xi = x - a_1 t - a_2 t^2 - a_3 t^3, \tag{8.5}$$

where a_1, a_2, and a_3 are constants, we can convert (8.4) into an ordinary differential equation,

$$\left(-a_1 - 2a_2 t - 3a_3 t^2\right)u' + t^2 u' + stuu' + rtu''' = 0, \tag{8.6}$$

where prime denotes the derivation with respect to ξ.

In order to eliminate the time-dependent coefficients in (8.6), we set

$$a_1 = 0, \qquad a_3 = \frac{1}{3}. \tag{8.7}$$

Equation (8.6) becomes

$$-2a_2 u' + suu' + ru''' = 0. \tag{8.8}$$

We rewrite (8.8) in the form

$$u' - \frac{s}{2a_2} uu' - \frac{r}{2a_2} u''' = 0. \tag{8.9}$$

Equation (8.9) can be effectively solved by the exp-function method (see Section 5.9). Hereby we consider a special case

$$\frac{s}{2a_2} = 6, \qquad \frac{r}{2a_2} = -1. \tag{8.10}$$

Equation (8.9) becomes the standard KdV equation, which admits the following solitary solution:

$$u = -\frac{A}{2}\operatorname{sech}^2\sqrt{\frac{A}{4}}\xi = -\frac{A}{2}\operatorname{sech}^2\sqrt{\frac{A}{4}}\left(x - a_2 t^2 - \frac{1}{3}t^3\right). \tag{8.11}$$

Acknowledgments

This work is supported by PAPD (a project funded by the Priority Academic Program Development of Jiangsu Higher Education Institutions), National Natural Science Foundation of China under Grants nos. 11061028 and 50806011, and Natural Science Foundation of Yunnan Province under Grant no. 2010CD086.

References

[1] J. S. Russell, "Report on waves," Tech. Rep., 1844, in *Proceedings of the14th Meeting of the British Association for the Advancement of Science*.

[2] N. J. Zabusky and M. D. Kruskal, "Interaction of "solitons" in a collisionless plasma and the recurrence of initial states," *Physical Review Letters*, vol. 15, no. 6, pp. 240–243, 1965.

[3] D. J. Korteweg and G. de Vires, "On the change of form of long waves advancing in a rectangular channel, and a new type of long stationary wave," *Philosophical Magazine*, vol. 539, pp. 422–443, 1895.

[4] C. S. Gardner, J. M. Greene, M. D. Kruskal, and R. M. Miura, "Method for solving the Korteweg-deVries equation," *Physical Review Letters*, vol. 19, no. 19, pp. 1095–1097, 1967.

[5] J.-H. He, "Soliton perturbation," in *Encyclopedia of Complexity and Systems Science*, R. Meyers, Ed., vol. 9, pp. 8453–8457, Springer, New York, NY, USA, 2009.

[6] J.-H. He and S. D. Zhu, "Solitons and compactons," in *Encyclopedia of Complexity and Systems Science*, R. Meyers, Ed., vol. 9, pp. 8457–8464, Springer, New York, NY, USA, 2009.

[7] J.-H. He, "Some asymptotic methods for strongly nonlinear equations," *International Journal of Modern Physics B*, vol. 20, no. 10, pp. 1141–1199, 2006.

[8] J. H. He, "An elementary introduction to recently developed asymptotic methods and nanomechanics in textile engineering," *International Journal of Modern Physics B*, vol. 22, no. 21, pp. 3487–3578, 2008.

[9] J.-H. He, "New interpretation of homotopy perturbation method. Addendum: some asymptotic methods for strongly nonlinear equations," *International Journal of Modern Physics B*, vol. 20, no. 18, pp. 2561–2568, 2006.

[10] P. Rosenau and J. M. Hyman, "Compactons: solitons with finite wavelength," *Physical Review Letters*, vol. 70, no. 5, pp. 564–567, 1993.

[11] S. Walia, R. Weber, K. Latham et al., "Oscillatory thermopower waves based on Bi_2Te_3 films," *Advanced Functional Materials*, vol. 21, no. 11, pp. 2072–2079, 2011.

[12] J. S. Lin and L. M. Hildemann, "A nonsteady-state analytical model to predict gaseous emissions of volatile organic compounds from landfills," *Journal of Hazardous Materials*, vol. 40, no. 3, pp. 271–295, 1995.

[13] W. Okrasiński and Ł. Płociniczak, "A nonlinear mathematical model of the corneal shape," *Nonlinear Analysis: Real World Applications*, vol. 13, no. 3, pp. 1498–1505, 2012.

[14] J.-H. He, "A remark on a nonlinear mathematical model of the corneal shape," *Nonlinear Analysis: Real World Applications*, vol. 13, no. 6, pp. 2863–2865, 2012.

[15] R. L. Herman, "Exploring the connection between quasistationary and squared eigenfunction expansion techniques in soliton perturbation theory," *Nonlinear Analysis: Theory, Methods & Applications*, vol. 63, no. 5–7, pp. e2473–e2482, 2005.

[16] J.-H. He, *Non-perturbative methods for strongly nonlinear problems [Ph.D. thesis]*, de-Verlag im Internet GmbH, Berlin, Germany, 2006.

[17] J.-H. He, "Variational principles for some nonlinear partial differential equations with variable coefficients," *Chaos, Solitons & Fractals*, vol. 19, no. 4, pp. 847–851, 2004.

[18] J.-H. He, *Generalized Variational Principles in Fluids*, Science & Culture Publishing House of China, 2003.

[19] M. J. Lighthill and G. B. Whitham, "On kinematic waves. II. A theory of traffic flow on long crowded roads," *Proceedings of the Royal Society London Series A*, vol. 229, pp. 317–345, 1955.

[20] G. B. Whitham, *Linear and Nonlinear Waves*, John Wiley & Sons, New York, NY, USA, 1974.

[21] W. Zheng, "A fluid dynamics model for the low speed traffic systems," *Acta Mechanica Sinica*, vol. 26, no. 2, pp. 149–157, 1994 (Chinese).

[22] J.-H. He, "Variational approach to foam drainage equation," *Meccanica*, vol. 46, no. 6, pp. 1265–1266, 2011.

[23] J.-H. He, "Variational approach to impulsive differential equations using the semi-inverse method," *Zeitschrift für Naturforschung A*, vol. 66, pp. 632–634, 2011.

[24] J.-H. He, "Variational approach for nonlinear oscillators," *Chaos, Solitons & Fractals*, vol. 34, no. 5, pp. 1430–1439, 2007.

[25] H. M. Liu and J.-H. He, "Variational approach to chemical reaction," *Computers and Chemical Engineering*, vol. 28, no. 9, p. 1549, 2004.

[26] T. Öziş and A. Yıldırım, "Application of He's semi-inverse method to the nonlinear Schrödinger equation," *Computers & Mathematics with Applications*, vol. 54, no. 7-8, pp. 1039–1042, 2007.

[27] J. Zhang, "Variational approach to solitary wave solution of the generalized Zakharov equation," *Computers & Mathematics with Applications*, vol. 54, no. 7-8, pp. 1043–1046, 2007.

[28] W. Dittrich and M. Reuter, *Classical and Quantum Dynamics*, Advanced Texts in Physics, Springer, Berlin, Germany, 2nd edition, 1994.

[29] A. Carini and F. Genna, "Some variational formulations for continuum nonlinear dynamics," *Journal of the Mechanics and Physics of Solids*, vol. 46, no. 7, pp. 1253–1277, 1998.

[30] G.-L. Liu, "A vital innovation in Hamilton principle and its extension to initial-value problems," in *Proceedings of the 4th International Conference on Nonlinear Mechanics*, pp. 90–97, Shanghai University Press, Shanghai, China, August 2002.

[31] J.-H. He, "Hamiltonian approach to nonlinear oscillators," *Physics Letters A*, vol. 374, no. 23, pp. 2312–2314, 2010.

[32] J. H. He, T. Zhong, and L. Tang, "Hamiltonian approach to duffing-harmonic equation," *The International Journal of Nonlinear Sciences and Numerical Simulation*, vol. 11, supplement, pp. 43–46, 2010.

[33] J. H. He, "Preliminary report on the energy balance for nonlinear oscillations," *Mechanics Research Communications*, vol. 29, no. 2-3, pp. 107–111, 2002.

[34] G. A. Afrouzi, D. D. Ganji, and R. A. Talarposhti, "He's energy balance method for nonlinear oscillators with discontinuities," *International Journal of Nonlinear Sciences and Numerical Simulation*, vol. 10, no. 3, pp. 301–304, 2009.

[35] H. L. Zhang, Y. G. Xu, and J. R. Chang, "Application of he's energy balance method to a nonlinear oscillator with discontinuity," *International Journal of Nonlinear Sciences and Numerical Simulation*, vol. 10, no. 2, pp. 207–214, 2009.

[36] S. S. Ganji, D. D. Ganji, and S. Karimpour, "He's Energy balance and He's variational methods for nonlinear oscillations in engineering," *International Journal of Modern Physics B*, vol. 23, no. 3, pp. 461–471, 2009.

[37] A. Yildirim, Z. Saadatnia, H. Askari, Y. Khan, and M. KalamiYazdi, "Higher order approximate periodic solutions for nonlinear oscillators with the Hamiltonian approach," *Applied Mathematics Letters*, vol. 24, no. 12, pp. 2042–2051, 2011.

[38] L. Xu, "A Hamiltonian approach for a plasma physics problem," *Computers and Mathematics with Applications*, vol. 61, no. 8, pp. 1909–1911, 2011.

[39] L. Xu and J. H. He, "Determination of limit cycle by Hamiltonian approach for strongly nonlinear oscillators," *International Journal of Nonlinear Sciences and Numerical Simulation*, vol. 11, no. 12, pp. 1097–1101, 2010.

[40] S. Durmaz, S. Altay Demirbağ, and M. O. Kaya, "High order hamiltonian approach to nonlinear oscillators," *International Journal of Nonlinear Sciences and Numerical Simulation*, vol. 11, no. 8, pp. 565–570, 2010.

[41] I. Didenkulova, E. Pelinovsky, and A. Sergeeva, "Statistical characteristics of long waves nearshore," *Coastal Engineering*, vol. 58, no. 1, pp. 94–102, 2011.

[42] C. C. Lin, "A new variational principle for isenergetic flows," *Quarterly of Applied Mathematics*, vol. 9, pp. 421–423, 1952.

[43] K. Washizu, *Variational Methods in Elasticity and Plasticity*, Pergamon Press, Oxford, UK, 1982.

[44] C. C. Lin, "Hydrodynamics of helium II," in *Proceedings of the International School of Physics*, vol. 21, pp. 93–146, Academic Press, 1963.

[45] J. C. Tannehill, D. A. Anderson, and R. H. Pletcher, *Computational Fluid Mechanics and Heat Transfer*, Taylor & Francis, 1997.

[46] J. W. Herivel, "The derivation of the equations of motion of an ideal fluid by Hamilton's principle," *Mathematical Proceedings of the Cambridge Philosophical Society*, vol. 51, pp. 344–349, 1955.

[47] F. P. Bretherton, "A note on Hamilton's principle for perfect fluids," *The Journal of Fluid Mechanics*, vol. 44, pp. 19–31, 1970.

[48] A. Ecer and H. U. Akay, "Investigation of transonic flow in a cascade using the finite element method," *Journal of American Institute of Aeronautics and Astronautics*, vol. 19, no. 9, pp. 1174–1182, 1981.

[49] A. Ecer and H. U. Akay, "A finite element formulation for steady transonic euler equations," *AIAA Journal*, vol. 21, no. 3, pp. 343–350, 1983.

[50] V. Goncharov and V. Pavlov, "On the Hamiltonian approach: applications to geophysical flows," *Nonlinear Processes in Geophysics*, vol. 5, no. 4, pp. 219–240, 1998.

[51] D. D. Holm and V. Zeitlin, "Hamilton's principle for quasigeostrophic motion," *Physics of Fluids*, vol. 10, no. 4, pp. 800–806, 1998.

[52] F. Hiroki and F. Youhei, "Clebsch potentials in the variational principle for a perfect fluid," *Progress of Theoretical Physics*, vol. 124, no. 3, pp. 517–531, 2010.

[53] J. Larsson, "A practical form of Lagrange-Hamilton theory for ideal fluids and plasmas," *Journal of Plasma Physics*, vol. 69, no. 3, pp. 211–252, 2003.

[54] R. Salmon, "Hamiltonian fluid mechanics," *Annual Review of Fluid Mechanics*, vol. 20, pp. 225–256, 1988.

[55] R. L. Seliger and G. B. Whitham, "Variational principles in continuum mechanics," *Proceedings of the Royal Society A*, vol. 305, pp. 1–25, 1968.

[56] H. J. Wagner, "On the use of Clebsch potentials in the Lagrangian formulation of classical electrodynamics," *Physics Letters A*, vol. 292, no. 4-5, pp. 246–250, 2002.

[57] A. Clebsch, "Uber die integration der hydrodynamischen gleichungen," *Journal für die Reine und Angewandte Mathematik*, vol. 56, no. 1, pp. 1–10, 1859.

[58] J.-H. He and X.-H. Wu, "Variational iteration method: new development and applications," *Computers & Mathematics with Applications*, vol. 54, no. 7-8, pp. 881–894, 2007.

[59] J.-H. He, "Variational iteration method—some recent results and new interpretations," *Journal of Computational and Applied Mathematics*, vol. 207, no. 1, pp. 3–17, 2007.

[60] J.-H. He and X.-H. Wu, "Construction of solitary solution and compacton-like solution by variational iteration method," *Chaos, Solitons & Fractals*, vol. 29, no. 1, pp. 108–113, 2006.

[61] J.-H. He, "A short remark on fractional variational iteration method," *Physics Letters A*, vol. 375, no. 38, pp. 3362–3364, 2011.

[62] J. H. He, G. C. Wu, and F. Austin, "The variational iterational method which should be follow," *Nonlinear Science Letters A*, vol. 1, pp. 1–30, 2010.

[63] L. M. B. Assas, "Variational iteration method for solving coupled-KdV equations," *Chaos, Solitons & Fractals*, vol. 38, no. 4, pp. 1225–1228, 2008.

[64] Z. M. Odibat, "Construction of solitary solutions for nonlinear dispersive equations by variational iteration method," *Physics Letters A*, vol. 372, no. 22, pp. 4045–4052, 2008.

[65] E. Hesameddini and H. Latifizadeh, "Reconstruction of variational iteration algorithms using the laplace transform," *International Journal of Nonlinear Sciences and Numerical Simulation*, vol. 10, no. 11-12, pp. 1377–1382, 2009.

[66] C. S. Pande and K. P. Cooper, "On the analytical solution for self-similar grain size distributions in two dimensions," *Acta Materialia*, vol. 59, no. 3, pp. 955–961, 2011.

[67] J. H. He et al., *First Course of Functional Analysis*, Asian Academic Publisher, 2011.

[68] L.-H. Zhou and J. H. He, "The variational approach coupled with an ancient Chinese mathematical method to the relativistic oscillator," *Mathematical & Computational Applications*, vol. 15, no. 5, pp. 930–935, 2010.

[69] J.-H. He, "Recent development of the homotopy perturbation method," *Topological Methods in Nonlinear Analysis*, vol. 31, no. 2, pp. 205–209, 2008.

[70] J. H. He, "Application of homotopy perturbation method to nonlinear wave equations," *Chaos, Solitons & Fractals*, vol. 26, no. 3, pp. 695–700, 2005.

[71] J. H. He, "A note on the homotopy perturbationmethod," *Thermal Science*, vol. 14, no. 2, pp. 565–568, 2010.

[72] J. H. He, "Homotopy perturbation method with an auxiliary term," *Abstract and Applied Analysis*, vol. 2012, Article ID 857612, 7 pages, 2012.

[73] Y. Khan and S. T. Mohyud-Din, "Coupling of He's polynomials and Laplace transformation for MHD viscous flow over a stretching sheet," *International Journal of Nonlinear Sciences and Numerical Simulation*, vol. 11, no. 12, pp. 1103–1107, 2010.

[74] W. Zhang and X. Li, "Approximate damped oscillatory solutions for generalized KdV-Burgers equation and their error estimates," *Abstract and Applied Analysis*, vol. 2011, Article ID 807860, 26 pages, 2011.

[75] H. Khan, S. Islam, J. Ali, and I. Ali Shah, "Comparison of different analytic solutions to axisymmetric squeezing fluid flow between two infinite parallel plates with slip boundary conditions," *Abstract and Applied Analysis*, vol. 2012, Article ID 835268, 18 pages, 2012.

[76] Y.-Q. Jiang and J.-M. Zhu, "Solitary wave solutions for a coupled MKdV system using the homotopy perturbation method," *Topological Methods in Nonlinear Analysis*, vol. 31, no. 2, pp. 359–367, 2008.

[77] B.-G. Zhang, S.-Y. Li, and Z.-R. Liu, "Homotopy perturbation method for modified Camassa-Holm and Degasperis-Procesi equations," *Physics Letters A*, vol. 372, no. 11, pp. 1867–1872, 2008.

[78] S. M. Schmalholz, "A simple analytical solution for slab detachment," *Earth and Planetary Science Letters*, vol. 304, no. 1-2, pp. 45–54, 2011.

[79] J.-H. He, "Modified Lindstedt-Poincaré methods for some strongly non-linear oscillations. I. Expansion of a constant," *International Journal of Non-Linear Mechanics*, vol. 37, no. 2, pp. 309–314, 2002.

[80] J.-H. He, "Bookkeeping parameter in perturbation methods," *International Journal of Nonlinear Sciences and Numerical Simulation*, vol. 2, no. 3, pp. 257–264, 2001.

[81] D. H. Shou and J. H. He, "Application of parameter-expanding method to strongly nonlinear oscillators," *International Journal of Nonlinear Sciences and Numerical Simulation*, vol. 8, no. 1, pp. 121–124, 2007.

[82] L. Xu, "Determination of limit cycle by He's parameter-expanding method for strongly nonlinear oscillators," *Journal of Sound and Vibration*, vol. 302, no. 1-2, pp. 178–184, 2007.

[83] S. Q. Wang and J. H. He, "Nonlinear oscillator with discontinuity by parameter-expansion method," *Chaos, Solitons & Fractals*, vol. 35, no. 4, pp. 688–691, 2008.

[84] L. Xu, "Application of He's parameter-expansion method to an oscillation of a mass attached to a stretched elastic wire," *Physics Letters, Section A*, vol. 368, no. 3-4, pp. 259–262, 2007.

[85] M. T. Darvishi, A. Karami, and B.-C. Shin, "Application of He's parameter-expansion method for oscillators with smooth odd nonlinearities," *Physics Letters A*, vol. 372, no. 33, pp. 5381–5384, 2008.

[86] F. Ö. Zengin, M. O. Kaya, and S. A. Demirbağ, "Application of parameter-expansion method to nonlinear oscillators with discontinuities," *International Journal of Nonlinear Sciences and Numerical Simulation*, vol. 9, no. 3, pp. 267–270, 2008.

[87] J.-H. He and X.-H. Wu, "Exp-function method for nonlinear wave equations," *Chaos, Solitons & Fractals*, vol. 30, no. 3, pp. 700–708, 2006.

[88] X.-H. Wu and J.-H. He, "Solitary solutions, periodic solutions and compacton-like solutions using the Exp-function method," *Computers & Mathematics with Applications*, vol. 54, no. 7-8, pp. 966–986, 2007.

[89] X.-H. Wu and J.-H. He, "EXP-function method and its application to nonlinear equations," *Chaos, Solitons & Fractals*, vol. 38, no. 3, pp. 903–910, 2008.

[90] A. Bekir and A. Boz, "Exact solutions for a class of nonlinear partial differential equations using exp-function method," *International Journal of Nonlinear Sciences and Numerical Simulation*, vol. 8, no. 4, pp. 505–512, 2007.

[91] R. Mokhtari, "Variational iteration method for solving nonlinear differential-difference equations," *International Journal of Nonlinear Sciences and Numerical Simulation*, vol. 9, no. 1, pp. 19–23, 2008.

[92] A. Yildirim, "Exact solutions of nonlinear differential-difference equations by He's homotopy perturbation method," *International Journal of Nonlinear Sciences and Numerical Simulation*, vol. 9, no. 2, pp. 111–114, 2008.

[93] S. Zhang, "Application of Exp-function method to high-dimensional nonlinear evolution equation," *Chaos, Solitons & Fractals*, vol. 38, no. 1, pp. 270–276, 2008.

[94] X.-W. Zhou, "Exp-function method for solving Huxley equation," *Mathematical Problems in Engineering*, vol. 2008, Article ID 538489, 7 pages, 2008.

[95] X. W. Zhou, Y. X. Wen, and J. H. He, "Exp-function method to solve the nonlinear dispersive $K(m, n)$ equations," *The International Journal of Nonlinear Sciences and Numerical Simulation*, vol. 9, pp. 301–306, 2008.

[96] A.-M. Wazwaz, "Solitary wave solutions of the generalized shallow water wave (GSWW) equation by Hirota's method, tanh-coth method and Exp-function method," *Applied Mathematics and Computation*, vol. 202, no. 1, pp. 275–286, 2008.

[97] Z. L. Tao, "A note on the variational approach to the Benjamin-Bona Mahony equation using He's semi-inverse method," *Int. J. Comput. Math.*, vol. 87, pp. 1752–1754, 2010.

[98] A.-M. Wazwaz, "Multiple-soliton solutions for the KP equation by Hirota's bilinear method and by the tanh-coth method," *Applied Mathematics and Computation*, vol. 190, no. 1, pp. 633–640, 2007.

[99] M. Wang, X. Li, and J. Zhang, "The (G'/G)-expansion method and travelling wave solutions of nonlinear evolution equations in mathematical physics," *Physics Letters A*, vol. 372, no. 4, pp. 417–423, 2008.

[100] S. Zhang, "A generalized auxiliary equation method and its application to $(2 + 1)$-dimensional Korteweg-de Vries equations," *Computers & Mathematics with Applications*, vol. 54, no. 7-8, pp. 1028–1038, 2007.

[101] L. Geng and X. C. Cai, "He's frequency formulation for nonlinear oscillators," *European Journal of Physics*, vol. 28, no. 5, pp. 923–931, 2007.

[102] H. L. Zhang, "Application of He's frequency-amplitude formulation to an $x^{1/3}$ force nonlinear oscillator," *International Journal of Nonlinear Sciences and Numerical Simulation*, vol. 9, no. 3, pp. 297–300, 2008.
[103] L. Zhao, "Chinese mathematics for nonlinear oscillators," *Topological Methods in Nonlinear Analysis*, vol. 31, no. 2, pp. 383–387, 2008.
[104] J. Fan, "Application of He's frequency-amplitude formulation to the Duffing-harmonic oscillator," *Topological Methods in Nonlinear Analysis*, vol. 31, no. 2, pp. 389–394, 2008.
[105] J. H. He, "An improved amplitude-frequency formulation for nonlinear oscillators," *The International Journal of Nonlinear Sciences and Numerical Simulation*, vol. 9, pp. 211–212, 2008.
[106] J. H. He, "Comment on 'He's frequency formulation for nonlinear oscillators'," *European Journal of Physics*, vol. 29, no. 4, pp. L19–L22, 2008.
[107] J. H. He, "Application of ancient chinese mathematics to optimal problems," *Nonlinear Science Letters A*, vol. 2, no. 2, pp. 81–84, 2011.
[108] S. T. Qin and Y. Ge, "A novel approach to Markowitz Portfolio model without using lagrange multipliers," *The International Journal of Nonlinear Sciences and Numerical Simulation*, vol. 11, supplement, pp. s331–s334, 2010.
[109] X. S. Cha, *Optimization and Optimal Control*, Tsinghua University Press, Beijing, China, 1983.
[110] J.-H. He and Q. Yang, "Solitary wavenumber-frequency formulation using an ancient Chinese arithmetic," *International Journal of Modern Physics B*, vol. 24, no. 24, pp. 4747–4751, 2010.
[111] J.-H. He, "He Chengtian's inequality and its applications," *Applied Mathematics and Computation*, vol. 151, no. 3, pp. 887–891, 2004.
[112] J. H. He, "Max-min approach to nonlinear oscillators," *The International Journal of Nonlinear Sciences and Numerical Simulation*, vol. 9, pp. 207–210, 2008.
[113] Y. V. Kartashov, V. A. Vysloukh, A. Malomed Boris, and L. Torner, "Solitons in nonlinear lattices," *Reviews of Modern Physics*, vol. 83, no. 1, pp. 247–305, 2011.
[114] G. C. Wu, L. Zhao, and J. H. He, "Differential-difference model for textile engineering," *Chaos, Solitons & Fractals*, vol. 42, no. 1, pp. 352–354, 2009.
[115] S. D. Zhu, "Exp-function method for the Hybrid-Lattice system," *The International Journal of Nonlinear Sciences and Numerical Simulation*, vol. 8, pp. 461–464, 2007.
[116] S. D. Zhu, "Exp-function method for the discrete mKdV lattice," *International Journal of Nonlinear Sciences and Numerical Simulation*, vol. 8, no. 3, pp. 465–468, 2007.
[117] C.-Q. Dai and Y.-Y. Wang, "Exact travelling wave solutions of the discrete nonlinear Schrödinger equation and the hybrid lattice equation obtained via the exp-function method," *Physica Scripta*, vol. 78, no. 1, article 015013, 6 pages, 2008.
[118] R. Mokhtari, "Variational iteration method for solvingnonlinear differential-difference equations," *International Journal of Nonlinear Sciences and Numerical Simulation*, vol. 9, no. 1, pp. 19–23, 2008.
[119] A. Yildirim, "Exact solutions of nonlinear differential-difference equations by He's homotopy perturbation method," *International Journal of Nonlinear Sciences and Numerical Simulation*, vol. 9, no. 2, pp. 111–114, 2008.
[120] J. He, "Some new approaches to Duffing equation with strongly and high order nonlinearity. II. Parametrized perturbation technique," *Communications in Nonlinear Science & Numerical Simulation*, vol. 4, no. 1, pp. 81–83, 1999.
[121] J.-H. He, "A review on some new recently developed nonlinear analytical techniques," *International Journal of Nonlinear Sciences and Numerical Simulation*, vol. 1, no. 1, pp. 51–70, 2000.
[122] X. H. Ding and L. Zhang, "Applying He's parameterized perturbation method for solving differential-difference equation," *International Journal of Nonlinear Sciences and Numerical Simulation*, vol. 10, no. 9, pp. 1249–1252, 2009.
[123] J.-H. He, S. K. Elagan, and G.-C. Wu, "Solitary-solution formulation for differential-difference equations using an ancient chinese algorithm," *Abstract and Applied Analysis*, vol. 2012, Article ID 861438, 6 pages, 2012.
[124] J.-H. He, "Approximate analytical solution for seepage flow with fractional derivatives in porous media," *Computer Methods in Applied Mechanics and Engineering*, vol. 167, no. 1-2, pp. 57–68, 1998.
[125] G. E. Drăgănescu, "Application of a variational iteration method to linear and nonlinear viscoelastic models with fractional derivatives," *Journal of Mathematical Physics*, vol. 47, no. 8, article 082902, 9 pages, 2006.
[126] Z. M. Odibat and S. Momani, "Application of variational iteration method to nonlinear differential equations of fractional order," *International Journal of Nonlinear Sciences and Numerical Simulation*, vol. 7, no. 1, pp. 27–34, 2006.

[127] N. T. Shawagfeh, "Analytical approximate solutions for nonlinear fractional differential equations," *Applied Mathematics and Computation*, vol. 131, no. 2-3, pp. 517–529, 2002.

[128] N. Bildik and A. Konuralp, "The use of variational iteration method, differential transform method and adomian decomposition method for solving different types of nonlinear partial differential equations," *International Journal of Nonlinear Sciences and Numerical Simulation*, vol. 7, no. 1, pp. 65–70, 2006.

[129] A. Ghorbani and J. Saberi-Nadjafi, "He's homotopy perturbation method for calculating adomian polynomials," *International Journal of Nonlinear Sciences and Numerical Simulation*, vol. 8, no. 2, pp. 229–232, 2007.

[130] A. Ghorbani, "Beyond Adomian polynomials: He polynomials," *Chaos, Solitons & Fractals*, vol. 39, no. 3, pp. 1486–1492, 2009.

[131] S. Momani and Z. Odibat, "Homotopy perturbation method for nonlinear partial differential equations of fractional order," *Physics Letters A*, vol. 365, no. 5-6, pp. 345–350, 2007.

[132] J. H. He, "A generalized poincaré-invariant action with possible application in strings and E-infinity theory," *Chaos, Solitons & Fractals*, vol. 39, no. 4, pp. 1667–1670, 2009.

[133] S. Das, "Solution of fractional vibration equation by the variational iteration method and modified decomposition method," *International Journal of Nonlinear Sciences and Numerical Simulation*, vol. 9, no. 4, pp. 361–366, 2008.

[134] Z. Odibat and S. Momani, "Applications of variational iteration and homotopy perturbation methods to fractional evolution equations," *Topological Methods in Nonlinear Analysis*, vol. 31, no. 2, pp. 227–234, 2008.

[135] S. Momani, Z. Odibat, and I. Hashim, "Algorithms for nonlinear fractional partial differential equations: a selection of numerical methods," *Topological Methods in Nonlinear Analysis*, vol. 31, no. 2, pp. 211–226, 2008.

[136] Z. Odibat and S. Momani, "Applications of variational iteration and homotopy perturbation methods to fractional evolution equations," *Topological Methods in Nonlinear Analysis*, vol. 31, no. 2, pp. 227–234, 2008.

[137] Z. Z. Ganji, D. D. Ganji, H. Jafari, and M. Rostamian, "Application of the homotopy perturbation method to coupled system of partial differential equations with time fractional derivatives," *Topological Methods in Nonlinear Analysis*, vol. 31, no. 2, pp. 341–348, 2008.

[138] Z.-B. Li and J.-H. He, "Fractional complex transform for fractional differential equations," *Mathematical & Computational Applications*, vol. 15, no. 5, pp. 970–973, 2010.

[139] J. H. He, S. K. Elagan, and Z. B. Li, "Geometrical explanation of the fractional complex transform and derivative chain rule for fractional calculus," *Physics Letters A*, vol. 376, no. 4, pp. 257–259, 2012.

[140] G. Jumarie, "Fractional partial differential equations and modified Riemann-Liouville derivative new methods for solution," *Journal of Applied Mathematics & Computing*, vol. 24, no. 1-2, pp. 31–48, 2007.

[141] X. Yang, "Local fractional integral transforms," *Progress in Nonlinear Science*, vol. 4, pp. 1–225, 2011.

[142] W. Chen, X. D. Zhang, and D. Korošak, "Investigation on fractional and fractal derivative relaxation-oscillation models," *International Journal of Nonlinear Sciences and Numerical Simulation*, vol. 11, no. 1, pp. 3–9, 2010.

[143] G. C. Wu, "Variational iteration method for q-difference equations of second order," *Journal of Applied Mathematics*, vol. 2012, Article ID 102850, 2012, http://www.hindawi.com/journals/jam/2012/102850/.

2

Regularity of Global Attractor for the Reaction-Diffusion Equation

Hong Luo

College of Mathematics and Software Science, Sichuan Normal University, Sichuan, Chengdu 610066, China

Correspondence should be addressed to Hong Luo, lhscnu@hotmail.com

Academic Editor: Giovanni Galdi

By using an iteration procedure, regularity estimates for the linear semigroups, and a classical existence theorem of global attractor, we prove that the reaction-diffusion equation possesses a global attractor in Sobolev space H^k for all $k > 0$, which attracts any bounded subset of $H^k(\Omega)$ in the H^k-norm.

1. Introduction

This paper is concerned with the following initial-boundary problem of reaction-diffusion systems involving an unknown function $u = u(x,t)$:

$$\begin{aligned} &\frac{\partial u}{\partial t} = a\Delta u - g(u) \quad \text{in } \Omega \times (0,\infty), \\ &u = 0, \quad \text{in } \partial\Omega \times (0,\infty), \\ &u(x,0) = \varphi, \quad \text{in } \Omega, \end{aligned} \tag{1.1}$$

where $a > 0$ is a given constant. Δ is the Laplace operator. Ω denotes an open bounded set of R^n $(n = 1,2,3)$ with smooth boundary $\partial\Omega$. $g(s)$ is a polynomial on $s \in R^1$, which is given by

$$g(s) = \sum_{k=1}^{p} a_k s^k, \tag{1.2}$$

where p should be an odd number, that is, $p = 2m + 1$ $(m \geq 1)$, and

$$a_p > 0. \tag{1.3}$$

The reaction-diffusion systems (1.1) have been extensively studied during the last decades, one of the motivations being that such systems could account for phenomena occurring in living organisms. In 1952, Turing [1] proposed that a combination of chemical reaction and diffusion produces spatial patterns of chemical concentration, under certain conditions. Such patterns are of interest because they give a possible explanation for the development of pattern and form in developmental biology [2–4] and experimental chemical systems [5]. Schneider et al. [6–9] have studied existence of periodic travelling wave solutions and positive periodic solution of reaction-diffusion systems. In [10–17], asymptotic behaviour of the nonlinear reaction-diffusion equation, such as global attractors, inertial manifolds, and approximate inertial manifolds, has been studied.

The global asymptotical behaviors of solutions and existence of global attractors are important for the study of the dynamical properties of general nonlinear dissipative dynamical systems. So, many authors are interested in the existence of global attractors such as [12–19]. As for the reaction-diffusion equation (1.1), the existence of global solutions and global attractors in $L^2(\Omega)$ has been proved by Temam [16], Marion [18, 19], and Zhong et al. [17]. For convenience, we introduce the main results as follows.

Lemma 1.1. *Under the conditions* (1.2) *and* (1.3) *for* $\varphi \in H$, *the following three claims hold.*

(1) *Equation* (1.1) *has a unique global weak solution* $u \in C((0,\infty); H) \bigcap L^2((0,T); H_{1/2})$, *for* $T > 0$;

(2) *Equation* (1.1) *has a unique strong solution* $u \in C([0,T); H_{1/2}) \bigcap L^2((0,T); H_1)$ *for any* $T > 0$;

(3) *Equation* (1.1) *has a global attractor* $\mathcal{A} \subset H$, *which attracts any bounded set of* H *in the* H*-norm.*

Here the spaces H, $H_{1/2}$ *and* H_1 *are defined as follows:*

$$H = L^2(\Omega), \qquad H_{1/2} = H_0^1(\Omega), \qquad H_1 = H_0^1(\Omega) \bigcap H^2(\Omega). \tag{1.4}$$

In this paper, we shall use the regularity estimates for the linear semigroups, combining with the classical existence theorem of global attractors, to prove that the reaction-diffusion equation possesses, in any kth differentiable function spaces $H^k(\Omega)$, a global attractor, which attracts any bounded set of $H^k(\Omega)$ in H^k-norm. The basic idea is an iteration procedure, which is from recent books and papers [20–24].

2. Preliminaries

Let X and X_1 be two Banach spaces and $X_1 \subset X$ a compact and dense inclusion. Consider the abstract nonlinear evolution equation defined on X, given by

$$\begin{aligned} \frac{du}{dt} &= Lu + G(u), \\ u(x,0) &= u_0, \end{aligned} \tag{2.1}$$

where $u(t)$ is an unknown function, $L : X_1 \to X$ a linear operator, and $G : X_1 \to X$ a nonlinear operator.

A family of operators $S(t) : X \to X$ $(t \geq 0)$ is called a semigroup generated by (2.1) provided that $S(t)$ satisfies the properties:

(1) $S(t) : X \to X$ is a continuous mapping for any $t \geq 0$;

(2) $S(0) = \text{id} : X \to X$ the identity;

(3) $S(t+s) = S(t) \cdot S(s)$ for all $t, s \geq 0$, and the solution of (2.1) can be expressed as

$$u(t, u_0) = S(t)u_0. \tag{2.2}$$

Next, we introduce the concepts and definitions of invariant sets, global attractors, ω-limit sets for the semigroup $S(t)$.

Definition 2.1. Let $S(t)$ be a semigroup defined on X. A set $\Sigma \subset X$ is called an invariant set of $S(t)$ if $S(t)\Sigma = \Sigma$ for all $t \geq 0$. An invariant set Σ is an attractor of $S(t)$ if Σ is compact, and there exists a neighborhood $U \subset X$ of Σ such that for any $u_0 \in U$:

$$\inf_{v \in \Sigma} \|S(t)u_0 - v\|_X \longrightarrow 0, \quad \text{as } t \longrightarrow \infty. \tag{2.3}$$

In this case, we say that Σ attracts U. Especially, if Σ attracts any bounded set of X, Σ is called a global attractor of $S(t)$.

For a set $D \subset X$, we define the ω-limit set of D as follows:

$$\omega(D) = \bigcap_{s \geq 0} \overline{\bigcup_{t \geq s} S(t)D}, \tag{2.4}$$

where the closure is taken in the X-norm. The following Lemma 2.2 is the classical existence theorem of global attractor by Temam [16].

Lemma 2.2. *Let $S(t) : X \to X$ be the semigroup generated by* (2.1). *Assume that the following conditions hold:*

(1) *$S(t)$ has a bounded absorbing set $B \subset X$, that is, for any bounded set $A \subset X$ there exists a time $t_A \geq 0$ such that $S(t)u_0 \in B$ for all $u_0 \subset A$ and $t > t_A$;*

(2) *$S(t)$ is uniformly compact, that is, for any bounded set $U \subset X$ and some $T > 0$ sufficiently large, the set $\overline{\bigcup_{t \geq T} S(t)U}$ is compact in X.*

Then, that the ω-limit set $\mathcal{A} = \omega(B)$ of B is a global attractor of (2.1), and $\mathcal{A}$ is connected providing B is connected.

Note that we used to assume that the linear operator L in (2.1) is a sectorial operator which generates an analytic semigroup e^{tL}. It is known that there exists a constant $\lambda \geq 0$ such that $L - \lambda I$ generates the fractional power operators $\mathcal{L}^\alpha$ and fractional order spaces X_α for $\alpha \in R^1$, where $\mathcal{L} = -(L - \lambda I)$. Without loss of generality, we assume that L generates the fractional power operators $\mathcal{L}^\alpha$ and fractional order spaces X_α as follows:

$$\mathcal{L}^\alpha = (-L)^\alpha : X_\alpha \longrightarrow X, \quad \alpha \in R^1, \tag{2.5}$$

where $X_\alpha = D(\mathcal{L}^\alpha)$ is the domain of $\mathcal{L}^\alpha$. By the semigroup theory of linear operators (Pazy [25]), we know that $X_\beta \subset X_\alpha$ is a compact inclusion for any $\beta > \alpha$.

Thus, Lemma 2.2 can be equivalently expressed in the following Lemma 2.3 [24].

Lemma 2.3. *Let $u(t, u_0) = S(t)u_0$ ($u_0 \in X, t \geq 0$) be a solution of (2.1) and $S(t)$ the semigroup generated by (2.1). Let X_α be the fractional order space generated by L. Assume*

(1) *for some $\alpha \geq 0$, there is a bounded set $B \subset X_\alpha$; for any $u_0 \in X_\alpha$, there exists $t_{u_0} > 0$ such that*

$$u(t, u_0) \in B, \quad \forall t > t_{u_0}; \tag{2.6}$$

(2) *there is a $\beta > \alpha$, for any bounded set $U \subset X_\beta$ there are $T > 0$ and $C > 0$ such that*

$$\|u(t, u_0)\|_{X_\beta} \leq C, \quad \forall t > T, \ u_0 \in U. \tag{2.7}$$

Then (2.1) has a global attractor $\mathcal{A} \subset X_\alpha$, which attracts any bounded set of X_α in the X_α-norm.

For sectorial operators, we also have the following properties, which can be found in [25].

Lemma 2.4. *Let $L : X_1 \to X$ be a sectorial operator which generates an analytic semigroup $T(t) = e^{tL}$. If all eigenvalues λ of L satisfy $\operatorname{Re}\lambda < -\lambda_0$ for some real number $\lambda_0 > 0$, then for $\mathcal{L}^\alpha$ ($\mathcal{L} = -L$) we have*

(1) *$T(t) : X \to X_\alpha$ is bounded for all $\alpha \in R^1$ and $t > 0$;*

(2) *$T(t)\mathcal{L}^\alpha x = \mathcal{L}^\alpha T(t)x$ for all $x \in X_\alpha$;*

(3) *for each $t > 0$, $\mathcal{L}^\alpha T(t) : X \to X$ is bounded, and*

$$\|\mathcal{L}^\alpha T(t)\| \leq C_\alpha t^{-\alpha} e^{-\delta t}, \tag{2.8}$$

where some $\delta > 0$, $C_\alpha > 0$ is a constant only depending on α;

(4) *the X_α-norm can be defined by*

$$\|x\|_{X_\alpha} = \|\mathcal{L}^\alpha x\|_X; \tag{2.9}$$

(5) *if* $\mathcal{L}$ *is symmetric, for any* $\alpha, \beta \in R^1$, *we have*

$$\langle \mathcal{L}^{\alpha} u, v \rangle_H = \left\langle \mathcal{L}^{\alpha-\beta} u, \mathcal{L}^{\beta} v \right\rangle_H. \tag{2.10}$$

3. Main Results

Let H and H_1 be the spaces defined as in (1.4). We define the operators $L : H_1 \to H$ and $G : H_1 \to H$ by

$$\begin{aligned} Lu &= a\Delta u, \\ Gu &= -g(u), \end{aligned} \tag{3.1}$$

where $g(u)$ is the same as one of (1.2). Thus, the reaction-diffusion equation (1.1) can be written in the abstract form (2.1). It is well known that the linear operator $L : H_1 \to H$ given by (3.1) is a sectorial operator and $\mathcal{L} = -L$. The space H_1 is the same as (1.4), $H_{1/2}$ is given by $H_{1/2}$ = closure of H_1 in $H^1(\Omega)$, and $H_k = H^{2k}(\Omega) \cap H_1$ for $k \geq 1$.

The main result in this paper is given by the following theorem, which provides the existence of global attractors of the reaction-diffusion equation (1.1) in any kth order space H_k.

Theorem 3.1. *Let the function* g *be a polynomial of order* p

$$g(u) = \sum_{k=1}^{p} a_k u^k, \qquad p = 2m + 1 \ (m \geq 1,\ m \in N), \tag{3.2}$$

with leading coefficient

$$a_p > 0. \tag{3.3}$$

Assume $p = 3$ *for* $n = 3$. *Then, for any* $\alpha \geq 0$ (1.1) *has a global attractor* $\mathcal{A}$ *in* H_α, *and* $\mathcal{A}$ *attracts any bounded set of* H_α *in the* H_α*-norm.*

Proof. From Lemma 1.1, we know that the solution of system (1.1) is a weak solution for any $\varphi \in H$. Hence, the solution $u(t, \varphi)$ of system (1.1) can be written as

$$u(t, \varphi) = e^{tL}\varphi + \int_0^t e^{(t-\tau)L} G(u)\, d\tau. \tag{3.4}$$

By (3.1), we rewrite (3.4) as

$$u(t, \varphi) = e^{tL}\varphi - \int_0^t e^{(t-\tau)L} g(u)\, d\tau. \tag{3.5}$$

Next, according to Lemma 2.3, we prove Theorem 3.1 in the following six steps.

Step 1. We prove that for any bounded set $U \subset H_{1/2}$ there is a constant $C > 0$ such that the solution $u(t, \varphi)$ of system (1.1) is uniformly bounded by the constant C for any $\varphi \in U$ and $t \geq 0$. To do that, we firstly check that system (1.1) has a global Lyapunov function as follows:

$$F(u) = \int_\Omega \left(\frac{1}{2} |\nabla u|^2 + f(u) \right) dx, \tag{3.6}$$

where

$$f(z) = \int_0^z g(z) dz = \sum_{k=1}^{p} \frac{1}{k+1} a_k z^{k+1}. \tag{3.7}$$

In fact, if $u(t, \cdot)$ is a weak solution of system (1.1), we have

$$\frac{d}{dt} F(u(t, \varphi)) = \left\langle DF(u), \frac{du}{dt} \right\rangle_H. \tag{3.8}$$

By (3.1) and (3.6), we get

$$\frac{du}{dt} = Lu + G(u) = -DF(u). \tag{3.9}$$

Hence, it follows from (3.8) and (3.9) that

$$\frac{dF(u)}{dt} = \langle DF(u), -DF(u) \rangle_H = -\|DF(u)\|_H^2, \tag{3.10}$$

which implies that (3.6) is a Lyapunov function.

Integrating (3.10) from 0 to t gives

$$F(u(t, \varphi)) = -\int_0^t \|DF(u)\|_H^2 dt + F(\varphi). \tag{3.11}$$

Using (1.2) and (3.6), we have

$$F(u) = \int_\Omega \left(\frac{1}{2} |\nabla u|^2 + f(u) \right) dx = \int_\Omega \left(\frac{1}{2} |\nabla u|^2 + \frac{1}{p+1} a_p u^{p+1} + \sum_{k=1}^{p-1} \frac{1}{k+1} a_k u^{k+1} \right) dx$$

$$\geq \int_\Omega \left(\frac{1}{2} |\nabla u|^2 + \frac{1}{p+1} a_p u^{p+1} - \sum_{k=1}^{p-1} \frac{1}{k+1} |a_k| |u|^{k+1} \right) dx$$

$$\geq \int_\Omega \left(\frac{1}{2}|\nabla u|^2 + \frac{1}{p+1} a_p u^{p+1} - \sum_{k=1}^{p-1} \frac{1}{k+1} |a_k| \left(\varepsilon |u|^{p+1} + \varepsilon^{-(k+1)/(p-k)} \right) \right) dx$$

$$= \int_\Omega \left(\frac{1}{2}|\nabla u|^2 + \frac{1}{p+1} a_p u^{p+1} - \varepsilon \left(\sum_{k=1}^{p-1} \frac{1}{k+1} |a_k| \right) |u|^{p+1} - \sum_{k=1}^{p-1} \frac{1}{k+1} |a_k| \varepsilon^{-(k+1)/(p-k)} \right) dx. \tag{3.12}$$

Choosing ε such that $\varepsilon(\sum_{k=1}^{p-1}(1/(k+1))|a_k|) = (1/2(p+1))a_p$, and noting that p is an odd number, that is, $p = 2m+1$ $(m \geq 1)$, we get

$$\begin{aligned} F(u) &\geq \int_\Omega \left(\frac{1}{2}|\nabla u|^2 + \frac{1}{p+1} a_p u^{p+1} - \frac{1}{2(p+1)} a_p |u|^{p+1} - \sum_{k=1}^{p-1} \frac{1}{k+1} |a_k| \varepsilon^{-(k+1)/(p-k)} \right) dx \\ &= \int_\Omega \left(\frac{1}{2}|\nabla u|^2 + \frac{1}{2(p+1)} a_p |u|^{p+1} - \sum_{k=1}^{p-1} \frac{1}{k+1} |a_k| \varepsilon^{-(k+1)/(p-k)} \right) dx \\ &\geq C_1 \int_\Omega \left(|\nabla u|^2 + |u|^{p+1} \right) dx - C_2. \end{aligned} \tag{3.13}$$

Combining with (3.11) yields

$$\begin{gathered} C_1 \int_\Omega \left(|\nabla u|^2 + |u|^{p+1} \right) dx - C_2 \leq -\int_0^t \|DF(u)\|_H^2 dt + F(\varphi), \\ C_1 \int_\Omega \left(|\nabla u|^2 + |u|^{p+1} \right) dx + \int_0^t \|DF(u)\|_H^2 dt \leq F(\varphi) + C_2, \\ \int_\Omega \left(|\nabla u|^2 + |u|^{p+1} \right) dx \leq C, \quad \forall t \geq 0, \ \varphi \in U, \end{gathered} \tag{3.14}$$

which implies

$$\|u(t,\varphi)\|_{H_{1/2}} \leq C, \quad \forall t \geq 0, \ \varphi \in U \subset H_{1/2}. \tag{3.15}$$

where C_1, C_2, and C are positive constants. C only depends on φ.

Step 2. We prove that for any bounded set $U \subset H_\alpha$ $(1/2 \leq \alpha < 1)$ there exists $C > 0$ such that

$$\|u(t,\varphi)\|_{H_\alpha} \leq C, \quad \forall t \geq 0, \ \varphi \in U, \ \alpha < 1. \tag{3.16}$$

By $H_{1/2} \hookrightarrow L^{2p}(\Omega)$, we have

$$\|g(u)\|_H^2 = \int_\Omega |g(u)|^2 dx = \int_\Omega \left|\sum_{k=1}^{p} a_k u^k\right|^2 dx \le \int_\Omega \left[a_p|u|^p + \sum_{k=1}^{p-1}|a_k|\left(\varepsilon|u|^p + \varepsilon^{-k/(p-k)}\right)\right]^2 dx$$
$$\le C\left(\int_\Omega |u|^{2p} dx + 1\right) \le C\left(\|u\|_{H_{1/2}}^{2p} + 1\right), \tag{3.17}$$

which implies that $g : H_{1/2} \to H$ is bounded.

Hence, it follows from (2.9) and (3.5) that

$$\begin{aligned}
\|u(t,\varphi)\|_{H_\alpha} &= \left\|e^{tL}\varphi + \int_0^t e^{(t-\tau)L} g(u)d\tau\right\|_{H_\alpha} \le \|\varphi\|_{H_\alpha} + \int_0^t \left\|\mathcal{L}^\alpha e^{(t-\tau)L} g(u)\right\|_H d\tau \\
&\le \|\varphi\|_{H_\alpha} + \int_0^t \left\|\mathcal{L}^\alpha e^{(t-\tau)L}\right\| \|g(u)\|_H d\tau \\
&\le \|\varphi\|_{H_\alpha} + C\int_0^t \left\|\mathcal{L}^\alpha e^{(t-\tau)L}\right\| \left(\|u\|_{H_{1/2}}^{2p} + 1\right) d\tau \\
&\le \|\varphi\|_{H_\alpha} + C\int_0^t \tau^\beta e^{-\delta t} d\tau \le C, \quad \forall t \ge 0, \ \varphi \in U \subset H_\alpha,
\end{aligned} \tag{3.18}$$

where $\beta = \alpha$ $(0 < \beta < 1)$. Hence, (3.16) holds.

Step 3. We prove that for any bounded set $U \subset H_\alpha$ $(1 \le \alpha < 3/2)$ there exists $C > 0$ such that

$$\|u(t,\varphi)\|_{H_\alpha} \le C, \quad \forall t \ge 0, \ \varphi \in U \subset H_\alpha, \ \alpha < \frac{3}{2}. \tag{3.19}$$

In fact, by the embedding theorems of fractional order spaces [25]:

$$H_\alpha \hookrightarrow C^0(\Omega) \cap H^1(\Omega), \quad \alpha \ge \frac{3}{4}, \tag{3.20}$$

we have

$$\|g(u)\|_{H_{1/2}}^2 = \int_\Omega |\nabla g(u)|^2 dx = \int_\Omega \left|\nabla\left(\sum_{k=1}^{p} a_k u^k\right)\right|^2 dx \le \int_\Omega \left|\sum_{k=1}^{p} k a_k u^{(k-1)} \nabla u\right|^2 dx$$
$$\le \int_\Omega \left[p a_p |u|^{p-1} + \sum_{k=1}^{p-1} k|a_k|\left(\varepsilon|u|^{p-1} + \varepsilon^{-(k-1)/(p-k)}\right)\right]^2 |\nabla u|^2 dx$$

$$
\begin{aligned}
&\le C\int_{\Omega}\left(|u|^{2p-2}+1\right)|\nabla u|^2dx \\
&\le C\int_{\Omega}\left(\sup_{x\in\Omega}|u|^{2p-2}+1\right)|\nabla u|^2dx \\
&\le C\left(\|u\|_{H_\alpha}^{2p-2}+1\right)\|u\|_{H_{1/2}}^2 \le C\left(\|u\|_{H_\alpha}^{2p-2}+1\right)\|u\|_{H_\alpha}^2,
\end{aligned}
\tag{3.21}
$$

which implies

$$
g : H_\alpha \longrightarrow H_{1/2} \text{ is bounded for } \alpha \ge \frac{3}{4}. \tag{3.22}
$$

Therefore, it follows from (3.16) and (3.22) that

$$
\|g(u)\|_{H_{1/2}} < C, \quad \forall t \ge 0,\ \varphi \in U \subset H_\alpha,\ \frac{3}{4} \le \alpha < 1. \tag{3.23}
$$

Then, by using same method as that in Step 2, we get from (3.23) that

$$
\begin{aligned}
\|u(t,\varphi)\|_{H_\alpha} &= \left\|e^{tL}\varphi + \int_0^t e^{(t-\tau)L}g(u)d\tau\right\|_{H_\alpha} \le \|\varphi\|_{H_\alpha} + \int_0^t \left\|\mathcal{L}^\alpha e^{(t-\tau)L}g(u)\right\|_H d\tau \\
&\le \|\varphi\|_{H_\alpha} + \int_0^t \left\|\mathcal{L}^\alpha e^{(t-\tau)L}\right\| \|g(u)\|_H d\tau \\
&\le \|\varphi\|_{H_\alpha} + C\int_0^t \left\|\mathcal{L}^{\alpha-(1/2)} e^{(t-\tau)L}\right\| \|g(u)\|_{H_{1/2}} d\tau \\
&\le \|\varphi\|_{H_\alpha} + C\int_0^t \tau^{-\beta}e^{-\delta t}d\tau \le C, \quad \forall t \ge 0,\ \varphi \in U \subset H_\alpha,
\end{aligned}
\tag{3.24}
$$

where $\beta = \alpha - 1/2$ $(0 < \beta < 1)$. Hence, (3.19) holds.

Step 4. We prove that for any bounded set $U \subset H_\alpha$ $(3/2 \le \alpha < 2)$ there exists $C > 0$ such that

$$
\|u(t,\varphi)\|_{H_\alpha} \le C, \quad \forall t \ge 0,\ \varphi \in U \subset H_\alpha,\ \alpha < 2. \tag{3.25}
$$

In fact, by the embedding theorems of fractional order spaces [25]:

$$
H^2 \hookrightarrow W^{1,6} \hookrightarrow W^{1,4}, \qquad H_\alpha \hookrightarrow C^0(\Omega) \cap H^2(\Omega), \quad \alpha \ge 1, \tag{3.26}
$$

we have

$$
\begin{aligned}
\|g(u)\|_{H_1}^2 &= \int_\Omega |\Delta g(u)|^2 dx = \int_\Omega \left|\Delta\left(\sum_{k=1}^{p} a_k u^k\right)\right|^2 dx \\
&= \int_\Omega \left|\sum_{k=2}^{p}\left[k(k-1)a_k u^{(k-2)}(\nabla u)^2 + k a_k u^{k-1}\Delta u\right] + a_1\Delta u\right|^2 dx \\
&\le \int_\Omega \Big\{p(p-1)a_p|u|^{p-2}(\nabla u)^2 + p a_p |u|^{p-1}\Delta u \\
&\qquad + \sum_{k=2}^{p-1}\Big[k(k-1)a_k\left(\varepsilon|u|^{p-2} + \varepsilon^{-(k-2)/(p-k)}\right)(\nabla u)^2 \\
&\qquad\qquad + k a_k\left(\varepsilon|u|^{p-1} + \varepsilon^{-(k-1)/(p-k)}\right)\Delta u\Big] + a_1|\Delta u|\Big\}^2 dx \\
&\le C\int_\Omega \left(|u|^{p-2}|\nabla u|^2 + |\nabla u|^2 + |u|^{p-1}|\Delta u| + |\Delta u|\right)^2 dx \\
&\le C\int_\Omega \left(|u|^{2p-4}|\nabla u|^4 + |\nabla u|^4 + |u|^{2p-2}|\Delta u|^2 + |\Delta u|^2\right) dx \\
&\le C\int_\Omega \left(\sup_{x\in\Omega}|u|^{2p-4}|\nabla u|^4 + |\nabla u|^4 + \sup_{x\in\Omega}|u|^{2p-2}|\Delta u|^2 + |\Delta u|^2\right) dx \\
&\le C\left(|u|_{H_\alpha}^{2p-4}\|u\|_{W^{1,4}}^4 + \|u\|_{W^{1,4}}^4 + |u|_{H_\alpha}^{2p-2}\|u\|_{H^2}^2 + \|u\|_{H^2}^2\right) \\
&\le C\left(|u|_{H^\alpha}^{2p-4}\|u\|_{H^2}^4 + \|u\|_{H^2}^4 + |u|_{H_\alpha}^{2p-2}\|u\|_{H^\alpha}^2 + \|u\|_{H^\alpha}^2\right) \\
&\le C\left(|u|_{H^\alpha}^{2p} + \|u\|_{H^\alpha}^4 + \|u\|_{H^\alpha}^2\right),
\end{aligned} \tag{3.27}
$$

which implies that

$$g : H_\alpha \longrightarrow H_1 \text{ is bounded for } \alpha \ge 1. \tag{3.28}$$

Therefore, it follows from (3.19) and (3.28) that

$$\|g(u)\|_{H_1} < C, \quad \forall t \ge 0,\ \varphi \in U \subset H_\alpha,\ 1 \le \alpha < \frac{3}{2}. \tag{3.29}$$

Then, we get from (3.29) that

$$\begin{aligned}\|u(t,\varphi)\|_{H_\alpha} &= \left\|e^{tL}\varphi+\int_0^t e^{(t-\tau)L}g(u)d\tau\right\|_{H_\alpha} \le \|\varphi\|_{H_\alpha}+\int_0^t\left\|\mathcal{L}^\alpha e^{(t-\tau)L}g(u)\right\|_H d\tau\\ &\le \|\varphi\|_{H_\alpha}+\int_0^t\left\|\mathcal{L}^\alpha e^{(t-\tau)L}\right\|\|g(u)\|_H d\tau\\ &\le \|\varphi\|_{H_\alpha}+\int_0^t\left\|\mathcal{L}^{\alpha-1} e^{(t-\tau)L}\right\|\|g(u)\|_{H_1} d\tau\\ &\le \|\varphi\|_{H_\alpha}+C\int_0^t \tau^{-\beta}e^{-\delta t}d\tau \le C,\quad \forall t\ge 0,\ \varphi\in U\subset H_\alpha,\end{aligned} \tag{3.30}$$

where $\beta=\alpha-1$ $(0<\beta<1)$. Hence, (3.25) holds.

Step 5. We prove that for any bounded set $U\subset H_\alpha$ $(\alpha\le 0)$ there exists $C>0$ such that

$$\|u(t,\varphi)\|_{H_\alpha}\le C,\quad \forall t\ge 0,\ \varphi\in U\subset H_\alpha,\ \alpha\ge 0. \tag{3.31}$$

In fact, by the embedding theorems of fractional order spaces [25]:

$$H^3\hookrightarrow W^{2,6}\hookrightarrow W^{1,6},\qquad H^3\hookrightarrow W^{2,6}\hookrightarrow W^{2,4},\qquad H_\alpha\hookrightarrow C^0(\Omega)\cap H^3(\Omega),\quad \alpha\ge\frac{3}{2}, \tag{3.32}$$

we have

$$\begin{aligned}\|g(u)\|_{H_{3/2}}^2 &= \int_\Omega|\nabla\Delta g(u)|^2dx=\int_\Omega\left|\nabla\Delta\left(\sum_{k=1}^p a_ku^k\right)\right|^2dx\\ &=\int_\Omega\left(\sum_{k=2}^p\Big[k(k-1)(k-2)a_k|u|^{k-3}|\nabla u|^3+3k(k-1)a_k|u|^{k-2}|\nabla u||\Delta u|\right.\\ &\qquad\left.+ka_k|u|^{k-1}|\nabla u|\Big]+a_1|\nabla u||\Delta u|\right)^2dx\\ &\le\int_\Omega\Big\{p(p-1)(p-2)a_p|u|^{p-3}|\nabla u|^3+3p(p-1)a_p|u|^{p-2}|\nabla u||\Delta u|+pa_p|u|^{p-1}|\nabla u|\\ &\qquad+\sum_{k=3}^{p-1}\Big[k(k-1)(k-2)a_k\left(\varepsilon|u|^{p-3}+\varepsilon^{-(k-3)/(p-k)}\right)|\nabla u|^3\\ &\qquad\quad+3k(k-1)a_k\left(\varepsilon|u|^{p-2}+\varepsilon^{-(k-2)/(p-k)}\right)|\nabla u||\Delta u|\\ &\qquad\quad+ka_k\left(\varepsilon|u|^{p-1}+\varepsilon^{-(k-1)/(p-k)}\right)|\nabla\Delta u|\Big]\\ &\qquad+6a_2|\nabla u||\Delta u|+2a_2|u||\nabla u|+a_1|\nabla\Delta u|\Big\}^2dx\end{aligned}$$

$$\leq C\int_{\Omega}\Big(|u|^{p-3}|\nabla u|^3+|\nabla u|^3+|u|^{p-2}|\nabla u||\Delta u|+|\nabla u||\Delta u|$$

$$+|u|^{p-1}|\nabla\Delta u|+|u||\nabla u|+|\nabla\Delta u|\Big)^2dx$$

$$\leq C\int_{\Omega}\Big(|u|^{2p-6}|\nabla u|^6+|\nabla u|^6+|u|^{2p-4}|\nabla u|^2|\Delta u|^2+|\nabla u|^2|\Delta u|^2$$

$$+|u|^{2p-2}|\nabla\Delta u|^2+|u|^2|\nabla u|^2+|\nabla\Delta u|^2\Big)dx$$

$$\leq C\int_{\Omega}\left(\sup_{x\in\Omega}|u|^{2p-6}|\nabla u|^6+|\nabla u|^6+\sup_{x\in\Omega}|u|^{2p-4}|\nabla u|^2|\Delta u|^2+|\nabla u|^2|\Delta u|^2\right.$$

$$\left.+\sup_{x\in\Omega}|u|^{2p-2}|\nabla\Delta u|^2+|u|^2|\nabla u|^2+|\nabla\Delta u|^2\right)dx$$

$$\leq C\Big(\|u\|_{H_\alpha}^{2p-6}\|u\|_{W^{1,6}}^6+\|u\|_{W^{1,6}}^6+\|u\|_{H_\alpha}^{2p-4}\|u\|_{W^{1,2}}^2\|u\|_{W^{2,2}}^2+\|u\|_{W^{2,4}}^4$$

$$+\|u\|_{H_\alpha}^{2p-2}\|u\|_{W^{3,2}}^2+\|u\|^2\|u\|_{W^{1,2}}^2+\|u\|_{W^{3,2}}^2\Big)$$

$$\leq C\Big(\|u\|_{H_\alpha}^{2p-6}\|u\|_{H_\alpha}^6+\|u\|_{H_\alpha}^{2p-4}\|u\|_{W^{2,4}}^4+\|u\|_{H_\alpha}^{2p-2}|u|_{H^3}^2+\|u\|_{H^3}^6+\|u\|_{W^{2,4}}^4+\|u\|_{H^3}^2\Big)$$

$$\leq C\Big(\|u\|_{H_\alpha}^{2p}+\|u\|_{H_\alpha}^{2p-4}\|u\|_{H_\alpha}^4+\|u\|_{H_\alpha}^{2p-2}\|u\|_{H_\alpha}^2+\|u\|_{H_\alpha}^6+\|u\|_{H_\alpha}^4+\|u\|_{H_\alpha}^2\Big)$$

$$\leq C\Big(\|u\|_{H_\alpha}^{2p}+\|u\|_{H_\alpha}^6+\|u\|_{H_\alpha}^4+\|u\|_{H_\alpha}^2\Big), \tag{3.33}$$

which implies that

$$g: H_\alpha \longrightarrow H_{3/2} \text{ is bounded for } \alpha \geq \frac{3}{2}. \tag{3.34}$$

Therefore, it follows from (3.25) and (3.34) that

$$\|g(u)\|_{H_{3/2}} < C, \quad \forall t \geq 0,\ \varphi \in U \subset H_\alpha,\ \frac{3}{2} \leq \alpha < 2. \tag{3.35}$$

Then, we get from (3.35) that

$$\begin{aligned}
\|u(t,\varphi)\|_{H_\alpha} &= \left\| e^{tL}\varphi + \int_0^t e^{(t-\tau)L} g(u) d\tau \right\|_{H_\alpha} \leq \|\varphi\|_{H_\alpha} + \int_0^t \left\| \mathcal{L}^\alpha e^{(t-\tau)L} g(u) \right\|_H d\tau \\
&\leq \|\varphi\|_{H_\alpha} + C \int_0^t \left\| \mathcal{L}^{\alpha-(3/2)} e^{(t-\tau)L} \right\| \|g(u)\|_{H_{3/2}} d\tau \\
&\leq \|\varphi\|_{H_\alpha} + C \int_0^t \tau^{-\beta} e^{-\delta t} d\tau \leq C, \quad \forall t \geq 0,\ \varphi \in U \subset H_\alpha,
\end{aligned} \tag{3.36}$$

where $\beta = \alpha - 3/2$ $(0 < \beta < 1)$. Hence, (3.31) holds for $2 \leq \alpha < 5/2$.

By doing the same procedures as Steps 1–4, we can prove that (3.31) holds for all $\alpha \geq 0$.

Step 6. We show that for any $\alpha \geq 0$, system (1.1) has a bounded absorbing set in H_α. We first consider the case of $\alpha = 1/2$.

It is well known that the reaction-diffusion equation possesses a global attractor in H space, and the global attractor of this equation consists of equilibria with their stable and unstable manifolds. Thus, each trajectory has to converge to a critical point. From (3.31) and (3.10), we deduce that for any $\varphi \in H_{1/2}$ the solution $u(t,\varphi)$ of system (1.1) converges to a critical point of F. Hence, we only need to prove the following two properties:

(1) $F(u) \to \infty \Leftrightarrow \|u\|_{H_{1/2}} \to \infty$;

(2) the set $S = \{u \in H_{1/2} \mid DF(u) = 0\}$ is bounded.

Property (1) is obviously true, we now prove property (2) in the following. It is easy to check if $DF(u) = 0$, u is a solution of the following equation:

$$\begin{gathered}
a\Delta u - g(u) = 0, \\
u|_{\partial\Omega} = 0,
\end{gathered} \tag{3.37}$$

where $g(u)$ is given by (1.2). Taking the scalar product of (3.37) with u, then we derive that

$$\int_\Omega \left(|\nabla u|^2 + \sum_{k=1}^{2m+1} a_k u^{k+1} \right) dx = 0. \tag{3.38}$$

By (1.3) and (3.38), we have

$$\int_\Omega \left(|\nabla u|^2 + a_{2m+1} u^{2(m+1)} - \sum_{k=1}^{2m} |a_k| |u|^{k+1} \right) dx \leq 0. \tag{3.39}$$

Using Hölder inequality and the above inequality, we have

$$\int_\Omega \left(|\nabla u|^2 + u^{2(m+1)} \right) dx \leq C, \tag{3.40}$$

where $C > 0$ is a constant. Thus, property (2) is proved.

Now, we show that system (1.1) has a bounded absorbing set in H_α for any $\alpha \geq 1/2$, that is, for any bounded set $U \subset H_\alpha$ there are $T > 0$ and a constant $C > 0$ independent of φ such that

$$\|u(t,\varphi)\|_{H_\alpha} \leq C, \quad \forall t \geq T, \ \varphi \in U. \tag{3.41}$$

From the above discussion, we know that (3.41) holds as $\alpha = 1/2$. By (3.5), we have

$$u(t,\varphi) = e^{(t-T)L}u(T,\varphi) - \int_0^t e^{(t-\tau)L}g(u)d\tau. \tag{3.42}$$

Let $B \subset H_{1/2}$ be the bounded absorbing set of system (1.1), and $T_0 > 0$ such that

$$u(t,\varphi) \in B, \quad \forall t \geq T_0, \ \varphi \in U \subset H_\alpha \left(\alpha \geq \frac{1}{2}\right). \tag{3.43}$$

It is well known that

$$\left\|e^{tL}\right\| \leq Ce^{-t\lambda_1}, \tag{3.44}$$

where $\lambda_1 > 0$ is the first eigenvalue of the equation:

$$\begin{aligned} a\Delta u - g(u) &= 0, \\ u|_{\partial\Omega} &= 0. \end{aligned} \tag{3.45}$$

Hence, for any given $T > 0$ and $\varphi \in U \subset H_\alpha$ $(\alpha \geq 1/2)$, we have

$$\left\|e^{(t-\tau)L}u(t,\varphi)\right\|_{H_\alpha} = \left\|\mathcal{L}^\alpha e^{(t-\tau)L}u(t,\varphi)\right\|_H \longrightarrow 0, \quad \text{as } t \longrightarrow \infty. \tag{3.46}$$

From (3.42)-(3.43) and Lemma 2.4 for any $1/2 \leq \alpha < 1$, we get that

$$\begin{aligned} \|u(t,\varphi)\|_{H_\alpha} &\leq \left\|e^{(t-T_0)L}u(T_0,\varphi)\right\|_{H_\alpha} + \int_{T_0}^t \left\|\mathcal{L}^\alpha e^{(t-\tau)L}g(u)\right\|d\tau \\ &\leq \left\|e^{(t-T_0)L}u(T_0,\varphi)\right\|_{H_\alpha} + C\int_0^{t-T_0} \tau^{-\alpha}e^{-\lambda_1\tau}d\tau, \end{aligned} \tag{3.47}$$

where $C > 0$ is a constant independent of φ.

Then, we infer from (3.46) and (3.47) that (3.41) holds for all $1/2 \leq \alpha < 1$. By the iteration method, we have that (3.41) holds for all $\alpha \geq 1/2$.

Finally, this theorem follows from (3.31)–(3.41) and Lemma 2.3. The proof is completed. □

Acknowledgment

This work is funded by the National Natural Science Foundation of China (no. 11071177) and the NSF of Sichuan Education Department of China (no. 11ZA1102).

References

[1] A. M. Turing, "The chemical basis of morphogenesis," *Philosophical Transactions of the Royal Society B*, vol. 237, pp. 37–72, 1952.
[2] L. G. Harrison, *Kinetic Theory of Living Pattern*, Cambridge University Press, 1993.
[3] H. Meinhardt, *Models of Biological Pattern Formation*, Academic Press, 1982.
[4] J. D. Murray, *Mathematical Biology*, vol. 19 of *Biomathematics*, Springer, Berlin, Germany, 1989.
[5] P. De Kepper, V. Castets, E. Dulos, and J. Boissonade, "Turing-type chemical patterns in the chlorite-iodide-malonic acid reaction," *Physica D*, vol. 49, no. 1-2, pp. 161–169, 1991.
[6] K. R. Schneider, "A note on the existence of periodic travelling wave solutions with large periods in generalized reaction-diffusion systems," *Journal of Applied Mathematics and Physics*, vol. 34, no. 2, pp. 236–240, 1983.
[7] S. J. Ruuth, "Implicit-explicit methods for reaction-diffusion problems in pattern formation," *Journal of Mathematical Biology*, vol. 34, no. 2, pp. 148–176, 1995.
[8] Y. Liu, Z. Li, and Q. Ye, "The existence, uniqueness and stability of positive periodic solution for periodic reaction-diffusion system," *Acta Mathematicae Applicatae Sinica*, vol. 17, no. 1, pp. 1–13, 2001.
[9] T. Miura and P. K. Maini, "Speed of pattern appearance in reaction-diffusion models: implications in the pattern formation of limb bud mesenchyme cells," *Bulletin of Mathematical Biology*, vol. 66, no. 4, pp. 627–649, 2004.
[10] M. Marion, "Approximate inertial manifolds for reaction-diffusion equations in high space dimension," *Journal of Dynamics and Differential Equations*, vol. 1, no. 3, pp. 245–267, 1989.
[11] J. Mallet-Paret and G. R. Sell, "Inertial manifolds for reaction diffusion equations in higher space dimensions," *Journal of the American Mathematical Society*, vol. 1, no. 4, pp. 805–866, 1988.
[12] J. K. Hale, *Asymptotic Behavior of Dissipative Systems*, vol. 25 of *Mathematical Surveys and Monographs*, American Mathematical Society, Providence, RI, USA, 1988.
[13] S. Lu, H. Wu, and C. Zhong, "Attractors for nonautonomous 2D Navier-Stokes equations with normal external forces," *Discrete and Continuous Dynamical Systems. Series A*, vol. 13, no. 3, pp. 701–719, 2005.
[14] Q. Ma, S. Wang, and C. Zhong, "Necessary and sufficient conditions for the existence of global attractors for semigroups and applications," *Indiana University Mathematics Journal*, vol. 51, no. 6, pp. 1541–1559, 2002.
[15] C. Zhong, C. Sun, and M. Niu, "On the existence of global attractor for a class of infinite dimensional dissipative nonlinear dynamical systems," *Chinese Annals of Mathematics. Series B*, vol. 26, no. 3, pp. 393–400, 2005.
[16] R. Temam, *Infinite-Dimensional Dynamical Systems in Mechanics and Physics*, vol. 68 of *Applied Mathematical Sciences*, Springer, New York, NY, USA, 2nd edition, 1997.
[17] C. K. Zhong, M.-H. Yang, and C.-Y. Sun, "The existence of global attractors for the norm-to-weak continuous semigroup and application to the nonlinear reaction-diffusion equations," *Journal of Differential Equations*, vol. 223, no. 2, pp. 367–399, 2006.
[18] M. Marion, "Attractors for reaction-diffusion equations: existence and estimate of their dimension," *Applicable Analysis*, vol. 25, no. 1-2, pp. 101–147, 1987.
[19] M. Marion, "Finite-dimensional attractors associated with partly dissipative reaction-diffusion systems," *SIAM Journal on Mathematical Analysis*, vol. 20, no. 4, pp. 816–844, 1989.
[20] T. Ma and S. H. Wang, *Bifurcation Theory and Applications*, vol. 53 of *World Scientific Series on Nonlinear Science, Series A*, World Scientific, Singapore, 2005.
[21] T. Ma and S. H. Wang, *Phase Transition Dynamics in Nonlinear Sciences*, Springer, NewYork, NY, USA, 2012.
[22] T. Ma and S. H. Wang, *Stability and Bifurcation of Nonlinear Evolution Equations*, Academic Press, China, 2006.
[23] B. Nicolaenko, B. Scheurer, and R. Temam, "Some global dynamical properties of a class of pattern formation equations," *Communications in Partial Differential Equations*, vol. 14, no. 2, pp. 245–297, 1989.

[24] L. Song, Y. Zhang, and T. Ma, "Global attractor of the Cahn-Hilliard equation in H^k spaces," *Journal of Mathematical Analysis and Applications*, vol. 355, no. 1, pp. 53–62, 2009.

[25] A. Pazy, *Semigroups of Linear Pperators and Applications to Partial Differential Equations*, vol. 44 of *Applied Mathematical Sciences*, Springer, New York, NY, USA, 1983.

3

Fixed Point Results for Almost Generalized Cyclic (ψ, φ)-Weak Contractive Type Mappings with Applications

Mohamed Jleli,[1] Erdal Karapınar,[2] and Bessem Samet[1]

[1] *Department of Mathematics, King Saud University, Riyadh 11451, Saudi Arabia*
[2] *Department of Mathematics, Atılım University, 06836 İncek, Ankara, Turkey*

Correspondence should be addressed to Erdal Karapınar, erdalkarapinar@yahoo.com

Academic Editor: Alexander I. Domoshnitsky

We define a class of almost generalized cyclic (ψ, ϕ)-weak contractive mappings and discuss the existence and uniqueness of fixed points for such mappings. We present some examples to illustrate our results. Moreover, we state some applications of our main results in nonlinear integral equations.

1. Introduction

Fixed point theory is a crucial tool in the analysis of nonlinear problems. Banach contraction mapping principle [1] is the most known result in this direction: A self-mapping $T : X \to X$ on a complete metric space (X, d) has a unique fixed point if there exists $k \in [0, 1)$ such that $d(Tx, Ty) \leq kd(x, y)$ for all $x, y \in X$. In this theorem, a self-mapping T is necessarily continuous. Due to its importance to fixed point theory in nonlinear analysis, desired Banach fixed point theorem have been heavily investigated by many authors (see, e.g., [2–15]). One of the remarkable generalizations of the banach contraction mapping principle was reported by Kirk et al. [16] via cyclic contraction. A mapping $T : A \cup B \to A \cup B$ is called cyclic if $T(A) \subseteq B$ and $T(B) \subseteq A$, where A, B are nonempty subsets of a metric space (X, d). Moreover, T is called cyclic contraction if there exists $k \in (0, 1)$ such that $d(Tx, Ty) \leq kd(x, y)$ for all $x \in A$ and $y \in B$. Notice that although a contraction is continuous, cyclic contraction need not to be. This is one of the important gains of this theorem. In this paper, the authors also introduced the following notion.

Definition 1.1 (see [16]). Let X be a nonempty set, p be a positive integer, and $T : X \to X$ be a mapping. $X = \bigcup_{i=1}^{p} A_i$ is said to be a cyclic representation of X with respect to T if

(i) A_i, $i = 1, 2, \ldots, p$ are nonempty closed sets,

(ii) $T(A_1) \subseteq A_2, \ldots, T(A_{p-1}) \subseteq A_p, T(A_p) \subseteq A_1$.

Following the paper in [16], a number of fixed point theorems on cyclic representation of X with respect to a self-mapping T have appeared (see, e.g., [17–26]).

The concept of almost contractions were introduced by Berinde [27, 28]. It was shown in [27] that any strict contraction, the Kannan [12] and Zamfirescu [29] mappings, as well as a large class of quasicontractions, are all almost contractions. Almost contractions and its generalizations were further considered in several works like [7, 30–37]. Recently, Ćirić et al. [7] proved some fixed point results in ordered metric spaces using almost generalized contractive condition, which is given in the following definition.

Definition 1.2. Let $T : X \to X$ be a self-mapping on a metric space (X, d). It is said to satisfy almost generalized contractive condition if there exists $k \in (0, 1)$ and $L \geq 0$ such that

$$\begin{aligned} d(Tx, Ty) \leq k \max\left\{ d(x, y), d(x, Tx), d(y, Ty), \frac{d(x, Ty) + d(y, Tx)}{2} \right\} \\ + L \min\{d(x, Tx), d(y, Ty), d(x, Ty), d(y, Tx)\}, \end{aligned} \tag{1.1}$$

for all $x, y \in X$.

In this paper, we introduce a class of almost generalized cyclic (ψ, φ)-weak contractive mappings and we investigate the existence and uniqueness of fixed points for almost generalized cyclic (ψ, φ)-weak contractive type mappings. Our main result generalizes and improves some well-known theorems in the literature (see, e.g., [16, 18, 19, 24, 26]). We state some examples to illustrate our results. Furthermore, we apply our main result to analyze the existence and uniqueness of solutions for a class of nonlinear integral equations.

2. Main Result

We start this section by defining two classes of real valued functions. Let Φ be the set of functions $\varphi : [0, \infty) \to [0, \infty)$ satisfying the following conditions:

(Φ_1) φ is lower semicontinuous;

(Φ_2) $\varphi^{-1}(\{0\}) = \{0\}$.

We denote by Ψ the set of all continuous functions $\psi : [0, \infty) \to [0, \infty)$.

Definition 2.1. Let (X,d) be a metric space. Let p be a positive integer and let $A_1, A_2, \ldots, A_p$ be nonempty subsets of X and $Y = \bigcup_{i=1}^{p} A_i$. Let $T : Y \to Y$ be a mapping such that

(I) $Y = \bigcup_{i=1}^{p} A_i$ is a cyclic representation of Y with respect to T,

(II) there exist $L \geq 0$, $\psi \in \Psi$ and $\varphi \in \Phi$ such that

$$\begin{aligned}\psi(d(Tx,Ty)) \leq \psi(M_T(x,y)) - \varphi(M_T(x,y)) \\ + L\min\{d(x,Tx), d(y,Ty), d(x,Ty), d(y,Tx)\},\end{aligned} \tag{2.1}$$

for all $(x,y) \in A_i \times A_{i+1}$, $i = 1,2,\ldots,p$ (with $A_{p+1} = A_1$), where

$$M_T(x,y) := \max\left\{d(x,y), d(Tx,x), d(Ty,y), \frac{d(x,Ty)+d(y,Tx)}{2}\right\}. \tag{2.2}$$

Then T is called an almost generalized cyclic (ψ,φ)-weak contractive mapping.

Remark 2.2. Taking in the above definition, $p = 1$, $A_1 = X$, $\psi(t) = t$, and $\varphi(t) = (1-k)t$, where $k \in (0,1)$ is a constant, we obtain an almost generalized contractive condition. Then any almost generalized contractive mapping is an almost generalized cyclic (ψ,φ)-weak contractive mapping.

Our main result is the following.

Theorem 2.3. *Let $\{A_i\}_{i=1}^{p}$ be nonempty closed subsets of a complete metric space (X,d) and $Y = \bigcup_{i=1}^{p} A_i$. Let $T : Y \to Y$ be an almost generalized cyclic (ψ,φ)-weak contractive mapping. Then T has a unique fixed point that belongs to $\bigcap_{i=1}^{p} A_i$.*

Proof. Let $x_0 \in A_1$ (such a point exists since $A_1 \neq \emptyset$). Define the sequence $\{x_n\}$ in X by

$$x_{n+1} = Tx_n, \quad n = 0,1,2,\ldots. \tag{2.3}$$

We will prove that

$$\lim_{n\to\infty} d(x_n, x_{n+1}) = 0. \tag{2.4}$$

If for some k, we have $x_{k+1} = x_k$, then (2.4) follows immediately. So, we can suppose that $d(x_n, x_{n+1}) > 0$ for all n. From the condition (I), we observe that for all n, there exists $i = i(n) \in \{1,2,\ldots,p\}$ such that $(x_n, x_{n+1}) \in A_i \times A_{i+1}$. Then, from the condition (II), we have

$$\psi(d(x_n, x_{n+1})) \leq \psi(M_T(x_{n-1}, x_n)) - \varphi(M_T(x_{n-1}, x_n)), \quad n = 1,2,\ldots. \tag{2.5}$$

On the other hand, we have

$$\begin{aligned} &M_T(x_{n-1}, x_n) \\ &= \max\left\{ d(x_{n-1}, x_n), d(Tx_{n-1}, x_{n-1}), d(Tx_n, x_n), \frac{d(x_{n-1}, Tx_n) + d(x_n, Tx_{n-1})}{2} \right\} \\ &= \max\left\{ d(x_{n-1}, x_n), d(x_n, x_{n-1}), d(x_{n+1}, x_n), \frac{d(x_{n-1}, x_{n+1})}{2} \right\} \\ &= \max\left\{ d(x_{n-1}, x_n), d(x_{n+1}, x_n), \frac{d(x_{n-1}, x_{n+1})}{2} \right\} \\ &= \max\{d(x_{n-1}, x_n), d(x_{n+1}, x_n)\}. \end{aligned} \tag{2.6}$$

Suppose that $M_T(x_{n-1}, x_n) = d(x_{n+1}, x_n)$. Using (2.5), we obtain

$$\psi(d(x_n, x_{n+1})) \leq \psi(d(x_n, x_{n+1})) - \varphi(d(x_n, x_{n+1})), \tag{2.7}$$

which implies that $\varphi(d(x_n, x_{n+1})) = 0$. From condition (Φ_2), we get that $d(x_n, x_{n+1}) = 0$, a contradiction with our assumption $d(x_n, x_{n+1}) > 0$ for all n. Thus, we have $M_T(x_{n-1}, x_n) = d(x_{n-1}, x_n)$, which implies that $\{d(x_n, x_{n+1})\}$ is a decreasing sequence of positive numbers. Then there exists $r \geq 0$ such that

$$\lim_{n\to\infty} d(x_n, x_{n+1}) = r. \tag{2.8}$$

Letting $n \to \infty$ in (2.5), using (2.8), the continuity of ψ and the lower semicontinuity of φ, we obtain

$$\psi(r) \leq \psi(r) - \varphi(r), \tag{2.9}$$

which implies that $\varphi(r) = 0$, that is, $r = 0$. Thus, we proved (2.4).

Now, we will prove that $\{x_n\}$ is a Cauchy sequence in (X, d). Suppose that $\{x_n\}$ is not a Cauchy sequence. Then there exists $\varepsilon > 0$ for which we can find two sequences of positive integers $\{m(k)\}$ and $\{n(k)\}$ such that for all positive integers k,

$$n(k) > m(k) > k, \qquad d(x_{m(k)}, x_{n(k)}) \geq \varepsilon, \qquad d(x_{m(k)}, x_{n(k)-1}) < \varepsilon. \tag{2.10}$$

Using (2.10) and the triangular inequality, we get

$$\begin{aligned} \varepsilon &\leq d(x_{n(k)}, x_{m(k)}) \\ &\leq d(x_{m(k)}, x_{n(k)-1}) + d(x_{n(k)-1}, x_{n(k)}) \\ &< \varepsilon + d(x_{n(k)}, x_{n(k)-1}). \end{aligned} \tag{2.11}$$

Thus, we have

$$\varepsilon \le d(x_{n(k)}, x_{m(k)}) < \varepsilon + d(x_{n(k)}, x_{n(k)-1}). \tag{2.12}$$

Letting $k \to \infty$ in the above inequality and using (2.4), we obtain

$$\lim_{k\to\infty} d(x_{n(k)}, x_{m(k)}) = \varepsilon^{+}. \tag{2.13}$$

On the other hand, for all k, there exists $j(k) \in \{1, \dots, p\}$ such that $n(k) - m(k) + j(k) \equiv 1[p]$. Then $x_{m(k)-j(k)}$ (for k large enough, $m(k) > j(k)$) and $x_{n(k)}$ lie in different adjacently labelled sets A_i and A_{i+1} for certain $i \in \{1, \dots, p\}$. Using (II), we obtain

$$\begin{aligned}\psi(d(Tx_{m(k)-j(k)}, Tx_{n(k)})) \le{}& \psi(M_T(x_{m(k)-j(k)}, x_{n(k)})) - \varphi(M_T(x_{m(k)-j(k)}, x_{n(k)}))\\ &+ L\min\{d(x_{m(k)-j(k)}, x_{m(k)-j(k)+1}), d(x_{n(k)}, x_{n(k)+1}),\\ &d(x_{m(k)-j(k)}, x_{n(k)+1}), d(x_{n(k)}, x_{m(k)-j(k)+1})\}\end{aligned} \tag{2.14}$$

for all k, that is,

$$\begin{aligned}\psi(d(x_{m(k)-j(k)+1}, x_{n(k)+1})) \le{}& \psi(M_T(x_{m(k)-j(k)}, x_{n(k)})) - \varphi(M_T(x_{m(k)-j(k)}, x_{n(k)}))\\ &+ L\min\{d(x_{m(k)-j(k)}, x_{m(k)-j(k)+1}), d(x_{n(k)}, x_{n(k)+1}),\\ &d(x_{m(k)-j(k)}, x_{n(k)+1}), d(x_{n(k)}, x_{m(k)-j(k)+1})\},\end{aligned} \tag{2.15}$$

for all k. Now, we have

$$\begin{aligned}&M_T(x_{m(k)-j(k)}, x_{n(k)})\\ &\quad= \max\left\{d(x_{m(k)-j(k)}, x_{n(k)}), d(x_{m(k)-j(k)+1}, x_{m(k)-j(k)}), d(x_{n(k)+1}, x_{n(k)}),\right.\\ &\qquad\left.\frac{d(x_{m(k)-j(k)}, x_{n(k)+1}) + d(x_{n(k)}, x_{m(k)-j(k)+1})}{2}\right\}\end{aligned} \tag{2.16}$$

for all k. Using the triangular inequality, we get

$$\begin{aligned}&\left|d(x_{m(k)-j(k)}, x_{n(k)}) - d(x_{n(k)}, x_{m(k)})\right|\\ &\quad\le d(x_{m(k)-j(k)}, x_{m(k)})\\ &\quad\le \sum_{l=0}^{j(k)-1} d(x_{m(k)-j(k)+l}, x_{m(k)-j(k)+l+1})\\ &\quad\le \sum_{l=0}^{p-1} d(x_{m(k)-j(k)+l}, x_{m(k)-j(k)+l+1}) \longrightarrow 0, \quad \text{as } k \longrightarrow \infty \quad (\text{from } (2.4)),\end{aligned} \tag{2.17}$$

which implies from (2.13) that

$$\lim_{k\to\infty} d\left(x_{m(k)-j(k)}, x_{n(k)}\right) = \varepsilon. \tag{2.18}$$

Using (2.4), we have

$$\lim_{k\to\infty} d\left(x_{m(k)-j(k)+1}, x_{m(k)-j(k)}\right) = 0, \tag{2.19}$$

$$\lim_{k\to\infty} d\left(x_{n(k)+1}, x_{n(k)}\right) = 0. \tag{2.20}$$

Again, using the triangular inequality, we get

$$\left|d\left(x_{m(k)-j(k)}, x_{n(k)+1}\right) - d\left(x_{m(k)-j(k)}, x_{n(k)}\right)\right| \leq d\left(x_{n(k)}, x_{n(k)+1}\right). \tag{2.21}$$

Letting $k \to \infty$ in the above inequality, using (2.20) and (2.18), we get

$$\lim_{k\to\infty} d\left(x_{m(k)-j(k)}, x_{n(k)+1}\right) = \varepsilon. \tag{2.22}$$

Similarly, we have

$$\left|d\left(x_{n(k)}, x_{m(k)-j(k)+1}\right) - d\left(x_{m(k)-j(k)}, x_{n(k)}\right)\right| \leq d\left(x_{m(k)-j(k)}, x_{m(k)-j(k)+1}\right). \tag{2.23}$$

Letting $k \to \infty$, using (2.4) and (2.18), we obtain

$$\lim_{k\to\infty} d\left(x_{n(k)}, x_{m(k)-j(k)+1}\right) = \varepsilon. \tag{2.24}$$

Similarly, we have

$$\lim_{k\to\infty} d\left(x_{m(k)-j(k)+1}, x_{n(k)+1}\right) = \varepsilon. \tag{2.25}$$

Now, it follows from (2.18)–(2.24) that

$$\begin{gathered}
\lim_{k\to\infty} M_T\left(x_{m(k)-j(k)}, x_{n(k)}\right) = \max\left\{\varepsilon, 0, 0, \frac{\varepsilon+\varepsilon}{2}\right\} = \varepsilon, \\
\lim_{k\to\infty} \min\left\{d\left(x_{m(k)-j(k)}, x_{m(k)-j(k)+1}\right), d\left(x_{n(k)}, x_{n(k)+1}\right), d\left(x_{m(k)-j(k)}, x_{n(k)+1}\right),\right. \\
\left. d\left(x_{n(k)}, x_{m(k)-j(k)+1}\right)\right\} = \min\{0, 0, \varepsilon, \varepsilon\} = 0.
\end{gathered} \tag{2.26}$$

Letting $k \to \infty$ in (2.15), using (2.25), (2.26), the continuity of ψ and the lower semi-continuity of φ, we obtain

$$\psi(\varepsilon) \leq \psi(\varepsilon) - \varphi(\varepsilon), \tag{2.27}$$

which implies that $\varphi(\varepsilon) = 0$, that is, $\varepsilon = 0$, a contradiction with $\varepsilon > 0$. Then we deduce that $\{x_n\}$ is a Cauchy sequence in the metric space (X, d).

Since (X, d) is complete, there exists $x^* \in X$ such that

$$\lim_{n\to\infty} x_n = x^*. \tag{2.28}$$

We will prove that

$$x^* \in \bigcap_{i=1}^{p} A_i. \tag{2.29}$$

From condition (I), and since $x_0 \in A_1$, we have $\{x_{np}\}_{n\geq 0} \subseteq A_1$. Since A_1 is closed, from (2.28), we get that $x^* \in A_1$. Again, from the condition (I), we have $\{x_{np+1}\}_{n\geq 0} \subseteq A_2$. Since A_2 is closed, from (2.28), we get that $x^* \in A_2$. Continuing this process, we obtain (2.29).

Now, we will prove that x^* is a fixed point of T. Indeed, from (2.29), since for all n, there exists $i(n) \in \{1, 2, \ldots, p\}$ such that $x_n \in A_{i(n)}$, Applying (II) with $x = x^*$ and $y = x_n$, we obtain

$$\begin{aligned} \psi(d(Tx^*, x_{n+1})) &= \psi(d(Tx^*, Tx_n)) \\ &\leq \psi(M_T(x^*, x_n)) - \varphi(M_T(x^*, x_n)) \\ &\quad + L\min\{d(x^*, Tx^*), d(x_n, x_{n+1}), d(x^*, x_{n+1}), d(x_n, Tx^*)\}, \end{aligned} \tag{2.30}$$

for all n. On the other hand, we have

$$M_T(x^*, x_n) = \max\left\{d(x^*, x_n), d(x^*, Tx^*), d(x_n, x_{n+1}), \frac{d(x^*, x_{n+1}) + d(x_n, Tx^*)}{2}\right\}. \tag{2.31}$$

Using (2.28), we obtain that

$$\begin{gathered} \lim_{n\to\infty} M_T(x^*, x_n) = d(x^*, Tx^*), \\ \lim_{n\to\infty} \min\{d(x^*, Tx^*), d(x_n, x_{n+1}), d(x^*, x_{n+1}), d(x_n, Tx^*)\} = 0. \end{gathered} \tag{2.32}$$

Letting $n \to \infty$ in (2.30), using (2.32), the continuity of ψ and the lower semicontinuity of φ, we get

$$\psi(d(x^*, Tx^*)) \leq \psi(d(x^*, Tx^*)) - \varphi(d(x^*, Tx^*)), \tag{2.33}$$

which implies that $d(x^*, Tx^*) = 0$, that is, x^* is a fixed point of T.

Finally, we prove that x^* is the unique fixed point of T. Assume that y^* is another fixed point of T, that is, $Ty^* = y^*$. From the condition (I), this implies that $y^* \in \bigcap_{i=1}^{p} A_i$. Then we can apply (II) for $x = x^*$ and $y = y^*$. We obtain

$$\psi(d(x^*, y^*)) \leq \psi(M_T(x^*, y^*)) - \varphi(M_T(x^*, y^*)). \tag{2.34}$$

Since x^* and y^* are fixed points of T, we can show easily that $M_T(x^*, y^*) = d(x^*, y^*)$. Then we get

$$\psi(d(x^*, y^*)) \leq \psi(d(x^*, y^*)) - \varphi(d(x^*, y^*)), \tag{2.35}$$

which implies that $d(x^*, y^*) = 0$, that is, $x^* = y^*$. Thus, we proved the uniqueness of the fixed point. □

3. Consequences

In this section, we derive some fixed point theorems from our main result given by Theorem 2.3.

If we take $p = 1$ and $A_1 = X$ in Theorem 2.3, then we get immediately the following fixed point theorem.

Corollary 3.1. *Let (X, d) be a complete metric space and $T : X \to X$ satisfies the following condition: there exist $L \geq 0$, $\psi \in \Psi$ and $\varphi \in \Phi$ such that*

$$\begin{aligned} \psi(d(Tx, Ty)) \leq \psi(M_T(x, y)) - \varphi(M_T(x, y)) \\ + L \min\{d(x, Tx), d(y, Ty), d(x, Ty), d(y, Tx)\}, \end{aligned} \tag{3.1}$$

for all $x, y \in X$. Then T has a unique fixed point.

An immediate consequence of Corollary 3.1 is the following fixed point theorem (see Remark 2.2).

Corollary 3.2. *Let (X, d) be a complete metric space and $T : X \to X$ satisfies the following condition: there exist $L \geq 0$ and $k \in (0, 1)$ such that*

$$d(Tx, Ty) \leq k M_T(x, y) + L \min\{d(x, Tx), d(y, Ty), d(x, Ty), d(y, Tx)\}, \tag{3.2}$$

for all $x, y \in X$. Then T has a unique fixed point.

Remark 3.3. Taking $L = 0$ in Corollary 3.1, we obtain Theorem 2.2 in [8]. Moreover, in Corollary 3.1, it is not supposed that ψ is nondecreasing and $\psi^{-1}(\{0\}) = \{0\}$, as in [8].

Taking $\psi(t) = t$, $\varphi(t) = (1 - k)t$, with $k \in (0, 1)$, in Theorem 2.3, we derive the following result.

Corollary 3.4. *Let $\{A_i\}_{i=1}^{p}$ be nonempty closed subsets of a complete metric space (X,d) and suppose that $T : \bigcup_{i=1}^{p} A_i \to \bigcup_{i=1}^{p} A_i$ satisfies the following conditions (where $A_{p+1} = A_1$):*

(i) *$T(A_i) \subseteq A_{i+1}$ for $1 \le i \le p$;*

(ii) *there exists a constant $k \in (0,1)$ such that*

$$d(Tx,Ty) \le k\max\left\{d(x,y), d(Tx,x), d(Ty,y), \frac{d(x,Ty)+d(y,Tx)}{2}\right\} + L\min\{d(x,Tx), d(y,Ty), d(x,Ty), d(y,Tx)\}, \tag{3.3}$$

for all $(x,y) \in (A_i, A_{i+1})$, for $1 \le i \le p$.

Then T has a unique fixed point that belongs to $\bigcap_{i=1}^{p} A_i$.

The following fixed point theorems established in [16, 25] are immediate consequences of the above result.

Corollary 3.5. *Let $\{A_i\}_{i=1}^{p}$ be nonempty closed subsets of a complete metric space (X,d) and suppose that $T : \bigcup_{i=1}^{p} A_i \to \bigcup_{i=1}^{p} A_i$ satisfies the following conditions (where $A_{p+1} = A_1$):*

(i) *$T(A_i) \subseteq A_{i+1}$ for $1 \le i \le p$;*

(ii) *there exists a constant $k \in (0,1)$ such that*

$$d(Tx,Ty) \le kd(x,y), \tag{3.4}$$

for all $(x,y) \in (A_i, A_{i+1})$, for $1 \le i \le p$.

Then T has a unique fixed point that belongs to $\bigcap_{i=1}^{p} A_i$.

Corollary 3.6. *Let $\{A_i\}_{i=1}^{p}$ be nonempty closed subsets of a complete metric space (X,d) and suppose that $T : \bigcup_{i=1}^{p} A_i \to \bigcup_{i=1}^{p} A_i$ satisfies the following conditions (where $A_{p+1} = A_1$):*

(i) *$T(A_i) \subseteq A_{i+1}$ for $1 \le i \le p$;*

(ii) *there exists a constant $k \in (0,1)$ such that*

$$d(Tx,Ty) \le \frac{k}{2}[d(x,Ty) + d(y,Tx)], \tag{3.5}$$

for all $(x,y) \in (A_i, A_{i+1})$, for $1 \le i \le p$.

Then T has a unique fixed point that belongs to $\bigcap_{i=1}^{p} A_i$.

Corollary 3.7. *Let* $\{A_i\}_{i=1}^p$ *be nonempty closed subsets of a complete metric space* (X, d) *and suppose that* $T : \bigcup_{i=1}^p A_i \to \bigcup_{i=1}^p A_i$ *satisfies the following conditions (where* $A_{p+1} = A_1$*):*

(i) $T(A_i) \subseteq A_{i+1}$ *for* $1 \le i \le p$*;*

(ii) *there exists a constant* $k \in (0, 1)$ *such that*

$$d(Tx, Ty) \le k \max\{d(x, Tx), d(y, Ty)\}, \tag{3.6}$$

for all $(x, y) \in (A_i, A_{i+1})$*, for* $1 \le i \le p$*.*

Then T *has a unique fixed point that belongs to* $\bigcap_{i=1}^p A_i$*.*

Corollary 3.8. *Let* $\{A_i\}_{i=1}^p$ *be nonempty closed subsets of a complete metric space* (X, d) *and suppose that* $T : \bigcup_{i=1}^p A_i \to \bigcup_{i=1}^p A_i$ *satisfies the following conditions (where* $A_{p+1} = A_1$*):*

(i) $T(A_i) \subseteq A_{i+1}$ *for* $1 \le i \le p$*;*

(ii) *there exist* $a_1, a_2, a_3, a_4 > 0$ *with* $a_1 + a_2 + a_3 + a_4 < 1$ *such that*

$$d(Tx, Ty) \le a_1 d(x, y) + a_2 d(Tx, x) + a_3 d(Ty, y) + a_4 \left[\frac{d(x, Ty) + d(y, Tx)}{2}\right], \tag{3.7}$$

for all $(x, y) \in (A_i, A_{i+1})$*, for* $1 \le i \le p$*.*

Then T *has a unique fixed point that belongs to* $\bigcap_{i=1}^p A_i$*.*

Remark 3.9. Taking $p = 1$ and $A_1 = X$, we get

(i) from Corollary 3.5, the Banach contraction principle [1];

(ii) from Corollary 3.6, Kannan's fixed point theorem [12];

(iii) from Corollary 3.7, Bianchini's fixed point theorem [2] (see also [3]);

(iv) from Corollary 3.8, Hardy and Rogers fixed point theorem [9].

Now, we derive a fixed point result for cyclic mappings satisfying a contractive condition of integral type.

Denote by Λ the set of functions $\alpha : [0, \infty) \to [0, \infty)$ satisfying the following hypotheses:

(Λ_1) α is a Lebesgue integrable mapping on each compact subset of $[0, \infty)$;

(Λ_2) for any $\varepsilon > 0$, we have $\int_0^\varepsilon \alpha(s)\, ds > 0$.

We have the following result.

Corollary 3.10. *Let* $\{A_i\}_{i=1}^{p}$ *be nonempty closed subsets of a complete metric space* (X,d) *and suppose* $T:\bigcup_{i=1}^{p}A_i\to\bigcup_{i=1}^{p}A_i$ *satisfies the following conditions (where* $A_{p+1}=A_1$*):*

(i) $T(A_i)\subseteq A_{i+1}$ *for* $1\le i\le p$;

(ii) *there exist* $L\ge 0$, $\alpha,\beta\in\Lambda$ *such that*

$$\int_0^{d(Tx,Ty)}\alpha(s)ds\le\int_0^{M_T(x,y)}\alpha(s)\,ds-\int_0^{M_T(x,y)}\beta(s)\,ds\\ +L\min\{d(x,Tx),d(y,Ty),d(x,Ty),d(y,Tx)\},\tag{3.8}$$

for all $(x,y)\in(A_i,A_{i+1})$, *for* $1\le i\le p$.

Then T has a unique fixed point that belongs to $\bigcap_{i=1}^{p}A_i$.

Proof. It follows immediately from Theorem 2.3 by observing that the functions $\psi(t)=\int_0^t\alpha(s)ds$ and $\varphi(t)=\int_0^t\beta(s)ds$ belong to Ψ. □

Taking $L=0$ in Corollary 3.10, we obtain the following result.

Corollary 3.11. *Let* $\{A_i\}_{i=1}^{p}$ *be nonempty closed subsets of a complete metric space* (X,d) *and suppose* $T:\bigcup_{i=1}^{p}A_i\to\bigcup_{i=1}^{p}A_i$ *satisfies the following conditions (where* $A_{p+1}=A_1$*):*

(i) $T(A_i)\subseteq A_{i+1}$ *for* $1\le i\le p$;

(ii) *there exist* $\alpha,\beta\in\Lambda$ *such that*

$$\int_0^{d(Tx,Ty)}\alpha(s)ds\le\int_0^{M_T(x,y)}\alpha(s)ds-\int_0^{M_T(x,y)}\beta(s)ds,\tag{3.9}$$

for all $(x,y)\in(A_i,A_{i+1})$, *for* $1\le i\le p$.

Then T has a unique fixed point that belongs to $\bigcap_{i=1}^{p}A_i$.

4. Some Examples

In this section, we give some examples to illustrate our obtained results.

Example 4.1. Let $X=[0,1]$ be endowed with the standard metric $d(x,y)=|x-y|$ for all $x,y\in X$. Consider the closed subsets A_1 and A_2 defined by $A_1=[0,1/4]$ and $A_2=[1/4,1]$. Define the mapping $T:X\to X$ by

$$Tx=\begin{cases}\dfrac{1}{4} & \text{if } x\in[0,1),\\ 0 & \text{if } x=1.\end{cases}\tag{4.1}$$

Clearly, we have $T(A_1)\subset A_2$ and $T(A_2)\subset A_1$.

Now, let $(x,y) \in A_1 \times A_2$. We distinguish two cases.

Case 1. If $y = 1$. In this case, we have

$$\begin{aligned} d(Tx,Ty) &= |Tx - Ty| \\ &= \left|\frac{1}{4} - 0\right| \\ &= \frac{1}{4} = \frac{1}{2}\cdot\frac{1}{2} \\ &\leq \frac{1}{2}d(y,Ty) \\ &\leq \frac{1}{2}M_T(x,y) \\ &= M_T(x,y) - \frac{1}{2}M_T(x,y). \end{aligned} \tag{4.2}$$

Define the functions $\psi, \varphi : [0,\infty) \to [0,\infty)$ by

$$\psi(t) = t, \quad \varphi(t) = \frac{t}{2}, \quad \forall t \geq 0. \tag{4.3}$$

Then we have

$$\psi(d(Tx,Ty)) \leq \psi(M_T(x,y)) - \varphi(M_T(x,y)). \tag{4.4}$$

Case 2. If $y \neq 1$. In this case, we have $d(Tx,Ty) = 0$, so inequality (4.4) is satisfied.

Similarly, if $(x,y) \in A_2 \times A_1$, we can show that (4.4) is satisfied.

Thus, we checked that all conditions of Theorem 2.3 are satisfied (with $p = 2$). We deduce that T has a unique fixed point $x^* \in A_1 \cap A_2 = \{1/4\}$.

Example 4.2. Let $X = [-\pi,\pi]$ be endowed with the standard metric $d(x,y) = |x - y|$ for all $x, y \in X$. Consider the closed subsets A_1 and A_2 defined by $A_1 = [0,\pi]$ and $A_2 = [-\pi,0]$. Define the mapping $T : X \to X$ by

$$Tx = \begin{cases} -\frac{1}{3}x\left|\cos\left(\frac{1}{x}\right)\right| & \text{if } x \in [-\pi,0) \cup (0,\pi], \\ 0 & \text{if } x = 0. \end{cases} \tag{4.5}$$

Clearly, we have $T(A_1) \subset A_2$ and $T(A_2) \subset A_1$.

Now, let $(x, y) \in (A_1 \times A_2)$ with $x \neq 0$ and $y \neq 0$. We have

$$\begin{aligned} d(Tx, Ty) &= |Tx - Ty| \\ &= \left| -\frac{1}{3}x\left|\cos\left(\frac{1}{x}\right)\right| + \frac{1}{3}y\left|\cos\left(\frac{1}{y}\right)\right|\right| \\ &= \frac{1}{3}\left||x|\left|\cos\left(\frac{1}{x}\right)\right| + |y|\left|\cos\left(\frac{1}{y}\right)\right|\right| \\ &\leq \frac{1}{3}(|x| + |y|). \end{aligned} \tag{4.6}$$

On the other hand, we have

$$\begin{gathered} |x| = x \leq x + \frac{1}{3}x\left|\cos\left(\frac{1}{x}\right)\right| = \left|x + \frac{1}{3}x\left|\cos\left(\frac{1}{x}\right)\right|\right| = d(x, Tx), \\ |y| = -y \leq -y + \frac{1}{3}\left|y\cos\left(\frac{1}{y}\right)\right| = -y - \frac{1}{3}y\left|\cos\left(\frac{1}{y}\right)\right| \leq \left|y + \frac{1}{3}y\left|\cos\left(\frac{1}{y}\right)\right|\right| = d(y, Ty). \end{gathered} \tag{4.7}$$

Then we have

$$d(Tx, Ty) \leq \frac{2}{3}\max\{d(x, Tx), d(y, Ty)\} \leq \frac{2}{3}M_T(x, y). \tag{4.8}$$

Consider the functions $\psi, \varphi : [0, \infty) \to [0, \infty)$ defined by

$$\psi(t) = t, \quad \varphi(t) = \frac{t}{3}, \quad \forall t \geq 0. \tag{4.9}$$

We have

$$\psi(d(Tx, Ty)) \leq \psi(M_T(x, y)) - \varphi(M_T(x, y)). \tag{4.10}$$

Moreover, we can show that the above inequality holds if $x = 0$ or $y = 0$.

Now, all conditions of Theorem 2.3 are satisfied (with $p = 2$), we deduce that T has a unique fixed point $x^* \in A_1 \cap A_2 = \{0\}$.

5. An Application

In this section, we apply the result given by Theorem 2.3 to study the existence and uniqueness of solutions to a class of nonlinear integral equations.

We consider the nonlinear integral equation

$$u(t) = \int_0^1 k(t, s, u(s))ds, \quad \forall t \in [0, 1], \tag{5.1}$$

where $k : [0, 1] \times [0, 1] \times \mathbb{R} \to \mathbb{R}$ is a continuous function.

Let $X = C([0,1])$ be the set of real continuous functions on $[0,1]$. We endow X with the following standard metric:

$$d_\infty(u,v) = \max_{t\in[0,1]}|u(t) - v(t)|, \quad \forall u,v \in X. \tag{5.2}$$

It is well known that (X, d_∞) is a complete metric space. Let $(\alpha,\beta) \in X^2$, let $(\alpha_0,\beta_0) \in \mathbb{R}^2$ such that

$$\alpha_0 \le \alpha \le \beta \le \beta_0. \tag{5.3}$$

We suppose that for all $t \in [0,1]$, we have

$$\alpha(t) \le \int_0^1 k(t,s,\beta(s))ds, \tag{5.4}$$

$$\beta(t) \ge \int_0^1 k(t,s,\alpha(s))ds. \tag{5.5}$$

We suppose that for all $t,s \in [0,1]$, $k(t,s,\cdot)$ is a decreasing function, that is,

$$x,y \in \mathbb{R}, \quad x \ge y \Longrightarrow k(t,s,x) \le k(t,s,y). \tag{5.6}$$

Finally, we suppose that for all $t,s \in [0,1]$, for all $x,y \in \mathbb{R}$ with $x \le \beta_0$ and $y \ge \alpha_0$ or $x \ge \alpha_0$ and $y \le \beta_0$,

$$\left|k(t,s,x) - k(t,s,y)\right| \le \xi(|x-y|), \tag{5.7}$$

where $\xi : [0,\infty) \to [0,\infty)$ is continuous nondecreasing and $\varphi : t \mapsto t - \xi(t)$ belongs to Φ. Now, define the set

$$\mathcal{W} = \{u \in C([0,1]) : \ \alpha \le u \le \beta\}. \tag{5.8}$$

We have the following result.

Theorem 5.1. *Under the assumptions* (5.3)–(5.7), *Problem* (5.1) *has one and only one solution* $u^* \in \mathcal{W}$.

Proof. Define the closed subsets of X, A_1, and A_2 by

$$\begin{aligned} A_1 &= \{u \in X : u \le \beta\}, \\ A_2 &= \{u \in X : u \ge \alpha\}. \end{aligned} \tag{5.9}$$

Define the mapping $T : X \to X$ by

$$Tu(t) = \int_0^1 k(t, s, u(s))ds, \quad \forall t \in [0, 1]. \tag{5.10}$$

We will prove that

$$T(A_1) \subseteq A_2, \qquad T(A_2) \subseteq A_1. \tag{5.11}$$

Let $u \in A_1$, that is,

$$u(s) \leq \beta(s), \quad \forall s \in [0, 1]. \tag{5.12}$$

Using condition (5.6), we obtain that

$$k(t, s, u(s)) \geq k(t, s, \beta(s)), \quad \forall t, s \in [0, 1]. \tag{5.13}$$

The above inequality with condition (5.4) imply that

$$\int_0^1 k(t, s, u(s))ds \geq \int_0^1 k(t, s, \beta(s))ds \geq \alpha(t), \tag{5.14}$$

for all $t \in [0, 1]$. Then we have $Tu \in A_2$.

Similarly, let $u \in A_2$, that is,

$$u(s) \geq \alpha(s), \quad \forall s \in [0, 1]. \tag{5.15}$$

Using condition (5.6), we obtain that

$$k(t, s, u(s)) \leq k(t, s, \alpha(s)), \qquad \forall t, s \in [0, 1]. \tag{5.16}$$

The above inequality with condition (5.5) imply that

$$\int_0^1 k(t, s, u(s))ds \leq \int_0^1 k(t, s, \alpha(s))ds \leq \beta(t), \tag{5.17}$$

for all $t \in [0, 1]$. Then we have $Tu \in A_1$. Finally, we deduce that (5.11) holds.

Now, let $(u, v) \in A_1 \times A_2$, that is, for all $t \in [0, 1]$,

$$u(t) \leq \beta(t), \qquad v(t) \geq \alpha(t). \tag{5.18}$$

This implies from condition (5.3) that for all $t \in [0, 1]$,

$$u(t) \leq \beta_0, \qquad v(t) \geq \alpha_0. \tag{5.19}$$

Now, using condition (5.7), we can write that for all $t \in [0,1]$, we have

$$\begin{aligned}|Tu - Tv|(t) &\leq \int_0^1 |k(t,s,u(s)) - k(t,s,v(s))|ds \\ &\leq \int_0^1 \xi(|u(s) - v(s)|)ds \\ &\leq \xi(d_\infty(u,v)) \quad (\text{since } \xi \text{ is nondecreasing}) \\ &\leq \xi(M_T(u,v)).\end{aligned} \tag{5.20}$$

This implies that

$$d_\infty(Tu,Tv) \leq \xi(M_T(u,v)) = M_T(u,v) - \varphi(M_T(u,v)). \tag{5.21}$$

Using the same technique, we can show that the above inequality holds also if we take $(u,v) \in A_2 \times A_1$.

Now, all the conditions of Theorem 2.3 are satisfied (with $\psi(t) = t$ and $L = 0$), we deduce that T has a unique fixed point $u^* \in A_1 \cap A_2 = \mathcal{W}$, that is, $u^* \in \mathcal{W}$ is the unique solution to (5.1). □

Acknowledgment

M. Jleli and B. Samet are supported by the Research Center, College of Science, King Saud University.

References

[1] S. Banach, "Sur les opérations dans les ensembles abstraits et leur application aux equations itegrales," *Fundamenta Mathematicae*, vol. 3, pp. 133–181, 1922.
[2] R. M. Tiberio Bianchini, "Su un problema di S. Reich riguardante la teoria dei punti fissi," vol. 5, pp. 103–108, 1972.
[3] S. K. Chatterjea, "Fixed-point theorems," *Comptes Rendus de l'Académie Bulgare des Sciences*, vol. 25, pp. 727–730, 1972.
[4] Y. J. Cho, P. P. Murthy, and G. Jungck, "A common fixed point theorem of Meir and Keeler type," *International Journal of Mathematics and Mathematical Sciences*, vol. 16, no. 4, pp. 669–674, 1993.
[5] Y. J. Cho, P. P. Murthy, and G. Jungck, "A theorem of Meir-Keeler type revisited," *International Journal of Mathematics and Mathematical Sciences*, vol. 23, no. 7, pp. 507–511, 2000.
[6] Y. J. E. Cho, R. Saadati, and G. Sadeghi, "Quasi-contractive mappings in modular metric spaces," *Journal of Applied Mathematics*, Article ID 907951, 5 pages, 2012.
[7] L. Ćirić, M. Abbas, R. Saadati, and N. Hussain, "Common fixed points of almost generalized contractive mappings in ordered metric spaces," *Applied Mathematics and Computation*, vol. 217, no. 12, pp. 5784–5789, 2011.
[8] D. Dorić, "Common fixed point for generalized (ψ, ϕ)-weak contractions," *Applied Mathematics Letters*, vol. 22, no. 12, pp. 1896–1900, 2009.
[9] G. E. Hardy and T. D. Rogers, "A generalization of a fixed point theorem of Reich," *Canadian Mathematical Bulletin*, vol. 16, pp. 201–206, 1973.
[10] G. Jungck, "Commuting mappings and fixed points," *The American Mathematical Monthly*, vol. 83, no. 4, pp. 261–263, 1976.

[11] G. Jungck, "Compatible mappings and common fixed points. II," *International Journal of Mathematics and Mathematical Sciences*, vol. 11, no. 2, pp. 285–288, 1988.
[12] R. Kannan, "Some results on fixed points," *Bulletin of the Calcutta Mathematical Society*, vol. 60, pp. 71–76, 1968.
[13] A. Meir and E. Keeler, "A theorem on contraction mappings," *Journal of Mathematical Analysis and Applications*, vol. 28, pp. 326–329, 1969.
[14] W. Sintunavarat and P. Kumam, "Common fixed point theorems for generalized J H-operator classes and invariant approximations," *Journal of Inequalities and Applications*, vol. 2011, article 67, 2011.
[15] W. Sintunavarat and P. Kumam, "Generalized common fixed point theorems in complex valued metric spaces and applications," *Journal of Inequalities and Applications*, vol. 2012, article 84, 2012.
[16] W. A. Kirk, P. S. Srinivasan, and P. Veeramani, "Fixed points for mappings satisfying cyclical contractive conditions," *Fixed Point Theory*, vol. 4, no. 1, pp. 79–89, 2003.
[17] R. P. Agarwal, M. A. Alghamdi, and N. Shahzad, "Fixed point theory for cyclic generalized contractions in partial metric spaces," *Fixed Point Theory and Applications*, vol. 2012, article 40, 2012.
[18] E. Karapınar, "Fixed point theory for cyclic weak ϕ-contraction," *Applied Mathematics Letters*, vol. 24, no. 6, pp. 822–825, 2011.
[19] E. Karapınar and K. Sadaranagni, "Fixed point theory for cyclic (ψ, ϕ)-contractions," *Fixed Point Theory and Applications*, vol. 2011, asrticle 69, 2011.
[20] E. Karapınar, N. Shobkolaei, S. Sedghi, and S. M. Vaezpour, "A common fixed point theorem for cyclic operators on partial metric spaces," *Filomat*, vol. 26, no. 2, pp. 407–414, 2012.
[21] E. Karapınar, "Best proximity points of cyclic mappings," *Applied Mathematics Letters*, vol. 25, no. 11, pp. 1761–1766, 2012.
[22] E. Karapınar and İ. M. Erhan, "Cyclic contraction and fixed point theorems," *Filomat*, vol. 26, no. 4, pp. 777–782, 2012.
[23] E. Karapınar, İ. M. Erhan, and A. Y. Ulus, "Fixed point theorem for cyclic maps on partial metric spaces," *Applied Mathematics & Information Sciences*, vol. 6, no. 1, pp. 239–244, 2012.
[24] M. Păcurar and I. A. Rus, "Fixed point theory for cyclic ϕ-contractions," *Nonlinear Analysis*, vol. 72, no. 3-4, pp. 1181–1187, 2010.
[25] M. A. Petric, "Some results concerning cyclical contractive mappings," *General Mathematics*, vol. 18, no. 4, pp. 213–226, 2010.
[26] I. A. Rus, "Cyclic representations and fixed points," *Annals of the Tiberiu Popoviciu Seminar of Functional Equations, Approximation and Convexity*, vol. 3, pp. 171–178, 2005.
[27] V. Berinde, "Approximating fixed points of weak contractions using the Picard iteration," *Nonlinear Analysis Forum*, vol. 9, no. 1, pp. 43–53, 2004.
[28] V. Berinde, "General constructive fixed point theorems for Cric-type almost contractions in metric spaces," *Carpathian Journal of Mathematics*, vol. 24, no. 2, pp. 10–19, 2008.
[29] T. Zamfirescu, "Fixed point theorems in metric spaces," *Archiv der Mathematik*, vol. 23, pp. 292–298, 1972.
[30] M. Abbas, G. V. R. Babu, and G. N. Alemayehu, "On common fixed points of weakly compatible mappings satisfying 'generalized condition (B)'," *Filomat*, vol. 25, no. 2, pp. 9–19, 2011.
[31] V. Berinde and M. Păcurar, "Fixed points and continuity of almost contractions," *Fixed Point Theory*, vol. 9, no. 1, pp. 23–34, 2008.
[32] B. Samet and C. Vetro, "Berinde mappings in orbitally complete metric spaces," *Chaos, Solitons & Fractals*, vol. 44, no. 12, pp. 1075–1079, 2011.
[33] T. Suzuki, "Fixed point theorems for Berinde mappings," *Bulletin of the Kyushu Institute of Technology. Pure and Applied Mathematics*, no. 58, pp. 13–19, 2011.
[34] W. Sintunavarat and P. Kumam, "Common fixed point theorem for hybrid generalized multi-valued contraction mappings," *Applied Mathematics Letters*, vol. 25, pp. 52–57, 2012.
[35] W. Sintunavarat and P. Kumam, "Weak condition for generalized multi-valued (f, α, β)-weak contraction mappings," *Applied Mathematics Letters*, vol. 24, no. 4, pp. 460–465, 2011.
[36] W. Sintunavarat and P. Kumam, "Common fixed point theorem for hybrid generalized multi-valued contraction mappings," *Applied Mathematics Letters*, vol. 25, no. 1, pp. 52–57, 2012.
[37] C. Mongkolkeha and P. Kumam, "Best proximity point theorems for generalized cyclic contractions in ordered metric spaces," *Journal of Optimization Theory and Applications*. In press.

4

Strong Convergence Theorems for Maximal Monotone Operators with Nonspreading Mappings in a Hilbert Space

Hongjie Liu, Junqing Wang, and Qiansheng Feng

School of Science, Tianjin Polytechnic University, Tianjin 300387, China

Correspondence should be addressed to Hongjie Liu, liuhongjie198891@126.com

Academic Editor: Yongfu Su

We prove the strong convergence theorems for finding a common element of the set of fixed points of a nonspreading mapping T and the solution sets of zero of a maximal monotone mapping and an α-inverse strongly monotone mapping in a Hilbert space. Manaka and Takahashi (2011) proved weak convergence theorems for maximal monotone operators with nonspreading mappings in a Hilbert space; there we introduced new iterative algorithms and got some strong convergence theorems for maximal monotone operators with nonspreading mappings in a Hilbert space.

1. Introduction

Let H be a real Hilbert space with inner product $\langle \cdot, \cdot \rangle$, and let C be a nonempty closed convex subset of H. We denote by $F(T)$ the set of fixed point of T. Then, a mapping $T : C \to C$ is said to be nonexpansive if $\|Tx - Ty\| \leq \|x - y\|$ for all $x, y \in C$. The mapping $T : C \to C$ is said to be firmly nonexpansive if $\|Tx - Ty\|^2 \leq \langle x - y, Tx - Ty \rangle$ for all $x, y \in C$; see, for instance, Browder [1] and Goebel and Kirk [2]. The mapping $T : C \to C$ is said to be firmly nonspreading [3] if

$$2\|Tx - Ty\|^2 \leq \|Tx - y\|^2 + \|x - Ty\|^2, \tag{1.1}$$

for all $x, y \in C$. Iemoto and Takahashi [4] proved that $T : C \to C$ is nonspreading if and only if

$$\|Tx - Ty\|^2 \leq \|x - y\|^2 + 2\langle x - Tx, y - Ty \rangle, \tag{1.2}$$

for all $x, y \in C$. It is not hard to know that a nonspreading mapping is deduced from a firmly nonexpansive mapping; see [5, 6], and a firmly nonexpansive mapping is a nonexpansive mapping.

Many studies have been done for structuring the fixed point of nonexpansive mapping T. In 1953, Mann [7] introduced the iteration as follows: a sequence $\{x_n\}$ defined by

$$x_{n+1} = \alpha_n x_n + (1 - \alpha_n) T x_n, \tag{1.3}$$

where the initial guess $x_1 \in C$ is arbitrary and $\{a_n\}$ is a real sequence in $[0,1]$. It is known that under appropriate settings, the sequence $\{x_n\}$ converges weakly to a fixed point of T. However, even in a Hilbert space, Mann iteration may fail to converge strongly, for example see [8].

Some attempts to construct iteration method guaranteeing the strong convergence have been made. For example, Halpern [9] proposed the following so-called Halpern iteration:

$$x_{n+1} = \alpha_n u + (1 - \alpha_n) T x_n, \tag{1.4}$$

where $u, x_1 \in C$ are arbitrary and $\{a_n\}$ is a real sequence in $[0,1]$ which satisfies $\alpha_n \to 0$, $\sum_{n=1}^{\infty} \alpha_n = \infty$ and $\sum_{n=1}^{\infty} |\alpha_n - \alpha_{n+1}| < \infty$. Then, $\{x_n\}$ converges strongly to a fixed point of T; see [9, 10].

In 1975, Baillon [11] first introduced the nonlinear ergodic theorem in Hilbert space as follows:

$$S_n x = \sum_{k=0}^{n-1} T^k x \tag{1.5}$$

converges weakly to a fixed point of T for some $x \in C$.

Recently, in the case when $T : C \to C$ is a nonexpansive mapping, $A : C \to H$ is an α-inverse strongly monotone mapping, and $B \in H \times H$ is a maximal monotone operator, Takahashi et al. [12] proved a strong convergence theorem for finding a point of $F(T) \cap (A + B)^{-1}(0)$, where $F(T)$ is the set of fixed points of T and $(A + B)^{-1}(0)$ is the set of zero points of $A + B$.

In 2011, Manaka and Takahashi [13] for finding a point of the set of fixed points of T and the set of zero points of $A + B$ in a Hilbert space, they introduced an iterative scheme as follows:

$$x_{n+1} = \beta_n x_n + (1 - \beta_n) T(J_{\lambda_n}(I - \lambda_n A) x_n), \tag{1.6}$$

where T is a nonspreading mapping, A is an α-inverse strongly monotone mapping, and B is a maximal monotone operator such that $J_\lambda = (I - \lambda B)^{-1}$; $\{\beta_n\}$ and $\{\lambda_n\}$ are sequences which satisfy $0 < c \leq \beta_n \leq d < 1$ and $0 < a \leq \lambda_n \leq b < 2\alpha$. Then they proved that $\{x_n\}$ converges weakly to a point $p = \lim_{n \to \infty} P_{F(T) \cap (A+B)^{-1}(0)} x_n$.

Motivated by above authors, we generalize and modify the iterative algorithms (1.5) and (1.6) for finding a common element of the set of fixed points of a nonspreading mapping T and the set of zero points of monotone operator $A + B$ (A is an α-inverse strongly monotone mapping, and B is a maximal monotone operator). First, we prove that the sequence generated by our iterative method is weak convergence under the property conditions. Then, we prove that the strong convergence in a Hilbert space. As expected, we get some weak and strong convergence theorems about the common element of the set of fixed points of a nonspreading mapping and the set of zero points of an α-inverse strongly monotone mapping and a maximal monotone operator in a Hilbert space.

2. Preliminaries

Let H be a real Hilbert space with inner product $\langle \cdot, \cdot \rangle$, and let C be a nonempty closed convex subset of H. A set-valued mapping $B : D(B) \subseteq H \to H$ is said to be monotone if for any $x, y \in D(B)$ and $x^* \in Bx$ and $y^* \in By$, it holds that

$$\langle x - y, x^* - y^* \rangle \geq 0. \tag{2.1}$$

A monotone operator B on H is said to be maximal if B has no monotone extension, that is, its graph is not properly contained in the graph of any other monotone operator on H. For a maximal monotone operator B on H and $r > 0$, we may define a single-valued operator $J_r = (I + rB)^{-1} : 2^H \to D(B)$, which is called the resolvent of B for $r > 0$. Let B be a maximal monotone operator on H, and let $B^{-1}(0) = \{x \in H : 0 \in Bx\}$. For a constant $\alpha > 0$, the mapping $A : C \to H$ is said to be an α-inverse strongly monotone if for any for all $x, y \in C$,

$$\langle x - y, Ax - Ay \rangle \geq \alpha \|Ax - Ay\|^2. \tag{2.2}$$

Remark 2.1. It is not hard to know that if A is an α-inverse strongly monotone mapping, then it is $1/\alpha$-Lipschitzian and hence uniformly continuous. Clearly, the class of monotone mappings include the class of an α-inverse strongly monotone mappings.

Remark 2.2. It is well known that if $T : C \to C$ is a nonexpansive mapping, then $I - T$ is $1/2$-inverse strongly monotone, where I is the identity mapping on H; see, for instance, [14]. It is known that the resolvent J_r is firmly nonexpansive and $B^{-1}(0) = F(J_r)$ for all $r > 0$.

For a single-valued mapping T, a point p is called a fixed point of T if $p = Tp$. For a multivalued mapping T, a point p is called a fixed point of T if $p \in Tp$. The set of fixed points of T is denoted by $F(T)$.

Let E be a uniformly convex real Banach space, K be a nonempty closed convex subset of E. A multivalued mapping $T : K \to CB(K)$ is said to be as follows.

(i) Contraction if there exists a constant $k \in [0, 1)$ such that

$$H(Tx, Ty) \leq k\|x - y\|, \quad \forall x, y \in K. \tag{2.3}$$

(ii) Nonexpansive if

$$H(Tx, Ty) \leq \|x - y\|, \quad \forall x, y \in K. \tag{2.4}$$

(iii) Quasinonexpansive if $F(T) \neq \emptyset$ and

$$H(Tx, Tp) \leq \|x - p\|, \quad \forall x \in K, \ \forall p \in F(T). \tag{2.5}$$

It is well known that every nonexpansive multivalued mapping T with $F(T) \neq \emptyset$ is multivalued quasinonexpansive. But there exist multivalued quasi-nonexpansive mappings that are not multivalued nonexpansive. It is clear that if T is a quasi-nonexpansive multivalued mapping, then $F(T)$ is closed.

A Banach space E is said to satisfy Opials condition if whenever $\{x_n\}$ is a sequence in E which converges weakly to x, then

$$\liminf_{n \to \infty} \|x_n - x\| < \liminf_{n \to \infty} \|x_n - y\|, \quad \forall y \in E, \ x \neq y. \tag{2.6}$$

Lemma 2.3 (Manaka and Takahashi [13]). *Let H be a real Hilbert space, and let C be a nonempty closed convex subset of H. Let $\alpha > 0$. Let A be an α-inverse strongly monotone mapping of C into H, and let B be a maximal monotone operator on H such that the domain of B is included in C. Let $J_\lambda = (I + \lambda B)^{-1}$ be the resolvent of B for any $\lambda > 0$. Then, the following hold*

(i) *if $u, v \in (A + B)^{-1}(0)$, then $Au = Av$;*

(ii) *for any $\lambda > 0$, $u \in (A + B)^{-1}(0)$ if and only if $u = J_\lambda(I - \lambda A)u$.*

Lemma 2.4 (Schu [15]). *Suppose that E is a uniformly convex Banach space and $0 < p \leq t_n \leq q < 1$ for all positive integers n. Also suppose that $\{x_n\}$ and $\{y_n\}$ are two sequences of E such that $\limsup_{n \to \infty} \|x_n\| \leq r$, $\limsup_{n \to \infty} \|y_n\| \leq r$, and $\lim_{n \to \infty} \|t_n x_n + (1 - t_n) y_n\| = r$ hold for some $r \geq 0$. Then, $\lim_{n \to \infty} \|x_n - y_n\| = 0$.*

Lemma 2.5 (Liu [16] and Xu [17]). *Let $\{a_n\}$ be a sequence of nonnegative real numbers satisfying the property as follows*

$$a_{n+1} \leq (1 - t_n) a_n + b_n + t_n c_n, \tag{2.7}$$

where $\{t_n\}$, $\{b_n\}$, and $\{c_n\}$ satisfy the restrictions as follows

(i) $\sum_{n=0}^{\infty} t_n = \infty$,

(ii) $\sum_{n=0}^{\infty} b_n < \infty$,

(iii) $\limsup_{n \to \infty} c_n \leq 0$.

Then, $\{a_n\}$ converges to zero as $n \to \infty$.

3. Strong Convergence Theorem

In this section, we prove the strong convergence theorems for finding a common element in common set of the fixed sets of a nonspreading mapping and the solution sets of zero of a maximal monotone operator and an α-inverse strongly monotone operator and in a Hilbert space.

Theorem 3.1. *Let C be a nonempty convex closed subset of a real Hilbert space H, let $A : C \to H$ be an α-inverse strongly monotone, let $B : D(B) \subseteq C \to 2^H$ be maximal monotone, let $J_\lambda = (I + \lambda B)^{-1}$ be the resolvent of B for any $\lambda > 0$, and let $T : C \to C$ be a nonspreading mapping. Assume that $F := F(T) \cap (A + B)^{-1}(0) \neq \emptyset$. We define*

$$\begin{aligned} x_1 &= x \in C, \text{ arbitrarily}, \\ z_n &= J_{\lambda_n}(I - \lambda_n A)x_n, \\ y_n &= \frac{1}{n}\sum_{k=1}^{n} T^k z_n, \\ x_{n+1} &= \alpha_n u + (1 - \alpha_n)y_n, \end{aligned} \tag{3.1}$$

where $\{\alpha_n\}$ is sequences in $[0,1]$ such that $\lim_{n\to\infty}\alpha_n = 0$, $\sum_{n=1}^{\infty}\alpha_n = \infty$. There exists a, b such that $0 < a \le \lambda_n \le b < 2\alpha$ for each $n \in N$. Then, $\{x_n\}$ converges strongly to Pu, and P is the metric projection of H onto F.

Proof. First, we prove that $\{x_n\}$ is bounded and $\lim_{n\to\infty}\|x_n - p\|$ exists for each $p \in F(T)$. In fact, from Lemma 2.3, we have $p = J_{\lambda_n}(I - \lambda_n A)p$, together with (3.1) and A is an α-inverse strongly monotone, we get that

$$\begin{aligned} \|z_n - p\|^2 &= \|J_{\lambda_n}(I - \lambda_n A)x_n - J_{\lambda_n}(I - \lambda_n A)p\|^2 \\ &\le \|(I - \lambda_n A)x_n - (I - \lambda_n A)p\|^2 \\ &= \|x_n - p\|^2 - 2\lambda_n\langle x_n - p, Ax_n - Ap\rangle + \lambda_n^2\|Ax_n - Ap\|^2 \\ &\le \|x_n - p\|^2 - 2\lambda_n\alpha\|Ax_n - Ap\|^2 + \lambda_n^2\|Ax_n - Ap\|^2 \\ &= \|x_n - p\|^2 - \lambda_n(2\alpha - \lambda_n)\|Ax_n - Ap\|^2 \\ &\le \|x_n - p\|^2. \end{aligned} \tag{3.2}$$

From the definition of y_n and T is nonspreading mapping, we obtain that

$$\begin{aligned} \|y_n - p\| &= \left\|\frac{1}{n}\sum_{k=0}^{n-1} T^k z_n - p\right\| \le \frac{1}{n}\sum_{k=0}^{n-1}\left\|T^k z_n - p\right\| \le \frac{1}{n}\sum_{k=0}^{n-1}\|z_n - p\| \\ &= \|z_n - p\| < \|x_n - p\|. \end{aligned} \tag{3.3}$$

Together with (3.1), we have that

$$\begin{aligned} \|x_{n+1} - p\| &= \|\alpha_n u + (1 - \alpha_n)y_n - p\| \\ &\le \alpha_n\|u - p\| + (1 - \alpha_n)\|y_n - p\| \\ &\le \alpha_n\|u - p\| + (1 - \alpha_n)\|x_n - p\|. \end{aligned} \tag{3.4}$$

Hence, we get that

$$\|x_{n+1} - p\| \leq \max\{\|u - p\|, \|x_n - p\|\}, \tag{3.5}$$

for all $n \in N$. This means that $\{x_n - p\}$ is bounded, so $\{x_n\}$ is bounded. From T is nonspreading, (3.3), and (3.2), we get that $\{y_n\}$, $\{z_n\}$, and $\{T^n z_n\}$ are all bounded.

Since $\{x_n\}$ is bounded, there exists a subsequence $\{x_{n_k}\}$ of $\{x_n\}$ such that $\lim_{k\to\infty}\|x_{n_k} - p\|$ exists. Since $\{x_{n_k}\}$ is bounded, there exists a subsequence $\{x_{n_{k_i}}\}$ of $\{x_{n_k}\}$ such that $x_{n_{k_i}} \rightharpoonup w \in C$ as $i \to \infty$. Now, we prove that $w \in F$. First, we prove that $w \in F(T)$. Since $\|x_{n+1} - y_n\| = \alpha_n\|u - y_n\|$, replacing n by n_{k_i}, we have $\|x_{n_{k_i}+1} - y_{n_{k_i}}\| = \alpha_{n_{k_i}}\|u - y_{n_{k_i}}\|$. Together with $\alpha_n \to 0$ and $\{y_n\}$ is bounded, we obtain that $\lim_{i\to\infty}\|x_{n_{k_i}+1} - y_{n_{k_i}}\| = 0$, so we have $y_{n_{k_i}} \rightharpoonup w$.

Let $n \in N$. Since T is nonspreading, we have that for all $y \in C$ and $k = 0, 1, 2, \ldots, n-1$,

$$\begin{aligned}\left\|T^{k+1}z_n - Ty\right\|^2 &\leq \left\|T^k z_n - y\right\|^2 + 2\left\langle T^k z_n - T^{k+1}z_n, y - Ty\right\rangle \\ &= \left\|T^k z_n - Ty\right\|^2 + \|Ty - y\|^2 + 2\left\langle T^k z_n - Ty, Ty - y\right\rangle \\ &\quad + 2\left\langle T^k z_n - T^{k+1}z_n, y - Ty\right\rangle.\end{aligned} \tag{3.6}$$

Summing these inequalities from $k = 0$ to $n - 1$ and dividing by n, we have

$$\frac{1}{n}\left(\|T^n z_n - Ty\|^2 - \|z_n - Ty\|^2\right) \leq \|Ty - y\|^2 + 2\langle y_n - Ty, Ty - y\rangle + \frac{2}{n}\langle z_n - T^n z_n, y - Ty\rangle. \tag{3.7}$$

Replacing n by n_{k_i}, we have

$$\begin{aligned}&\frac{1}{n_{k_i}}\left(\left\|T^{n_{k_i}}z_{n_{k_i}} - Ty\right\|^2 - \left\|z_{n_{k_i}} - Ty\right\|^2\right) \\ &\quad\leq \|Ty - y\|^2 + 2\left\langle y_{n_{k_i}} - Ty, Ty - y\right\rangle \\ &\qquad + \frac{2}{n_{k_i}}\left\langle z_{n_{k_i}} - T^{n_{k_i}}z_{n_{k_i}}, y - Ty\right\rangle.\end{aligned} \tag{3.8}$$

Since $\{z_n\}$ and $\{T^n z_n\}$ are bounded, we have that

$$0 \leq \|Ty - y\|^2 + 2\langle w - Ty, Ty - y\rangle \tag{3.9}$$

as $i \to \infty$. Putting $y = w$, we have

$$0 \leq \|Tw - w\|^2 + 2\langle w - Tw, Tw - w\rangle = -\|Tw - w\|^2. \tag{3.10}$$

Hence, $w \in F(T)$.

Next, we prove that $w \in (A+B)^{-1}(0)$. From (3.2) and (3.3) we have that

$$\begin{aligned}\|x_{n+1}-p\|^2 &\le \alpha_n\|u-p\|^2+(1-\alpha_n)\|y_n-p\|^2\\ &\le \alpha_n\|u-p\|^2+(1-\alpha_n)\|z_n-p\|^2\\ &\le \alpha_n\|u-p\|^2+(1-\alpha_n)\left(\|x_n-p\|^2-\lambda_n(2\alpha-\lambda_n)\|Ax_n-Ap\|^2\right)\\ &= \alpha_n\left(\|u-p\|^2-\|x_n-p\|^2\right)+\|x_n-p\|^2-(1-\alpha_n)\lambda_n(2\alpha-\lambda_n)\|Ax_n-Ap\|^2.\end{aligned} \tag{3.11}$$

We rewrite above inequality as follows:

$$(1-\alpha_n)\lambda_n(2\alpha-\lambda_n)\|Ax_n-Ap\|^2 \le \alpha_n\left(\|u-p\|^2-\|x_n-p\|^2\right)+\|x_n-p\|^2-\|x_{n+1}-p\|^2. \tag{3.12}$$

Replacing n by n_k, we have

$$\begin{aligned}&(1-\alpha_{n_k})\lambda_{n_k}(2\alpha-\lambda_{n_k})\|Ax_{n_k}-Ap\|^2\\ &\le \alpha_{n_k}\left(\|u-p\|^2-\|x_{n_k}-p\|^2\right)\\ &\quad+\|x_{n_k}-p\|^2-\|x_{n_k+1}-p\|^2.\end{aligned} \tag{3.13}$$

Together with $\lim_{n\to\infty}\alpha_n=0$, $0<a\le\lambda_n\le b<2\alpha$ and since $\lim_{k\to\infty}\|x_{n_k}-p\|$ exists, we obtain that

$$\lim_{k\to\infty}\|Ax_{n_k}-Ap\|=0. \tag{3.14}$$

Since J_{λ_n} is firmly nonexpansive, and from (3.2), we have that

$$\begin{aligned}\|z_n-p\|^2 &= \|J_{\lambda_n}(I-\lambda_nA)x_n-J_{\lambda_n}(I-\lambda_nA)p\|^2\\ &\le \langle z_n-p,(I-\lambda_nA)x_n-(I-\lambda_nA)p\rangle\\ &= \frac{1}{2}\Big\{\|z_n-p\|^2+\|(I-\lambda_nA)x_n-(I-\lambda_nA)p\|^2\\ &\qquad -\|z_n-p-(I-\lambda_nA)x_n+(I-\lambda_nA)p\|^2\Big\}\\ &\le \frac{1}{2}\Big\{\|z_n-p\|^2+\|x_n-p\|^2-\|z_n-p-(I-\lambda_nA)x_n+(I-\lambda_nA)p\|^2\Big\}\\ &= \frac{1}{2}\Big\{\|z_n-p\|^2+\|x_n-p\|^2-\|z_n-x_n\|^2-2\lambda_n\langle z_n-x_n,Ax_n-Ap\rangle-\lambda_n^2\|Ax_n-Ap\|^2\Big\}.\end{aligned} \tag{3.15}$$

This means that

$$\|z_n - p\|^2 \le \|x_n - p\|^2 - \|z_n - x_n\|^2 - 2\lambda_n\langle z_n - x_n, Ax_n - Ap\rangle - \lambda_n^2\|Ax_n - Ap\|^2. \tag{3.16}$$

Together with (3.1) and (3.3), we have

$$\begin{aligned}
\|x_{n+1} - p\|^2 &\le \alpha_n\|u - p\|^2 + (1 - \alpha_n)\|y_n - p\|^2 \\
&\le \alpha_n\|u - p\|^2 + (1 - \alpha_n)\|z_n - p\|^2 \\
&\le \alpha_n\|u - p\|^2 + (1 - \alpha_n) \\
&\quad \times \left\{\|x_n - p\|^2 - \|z_n - x_n\|^2 - 2\lambda_n\langle z_n - x_n, Ax_n - Ap\rangle - \lambda_n^2\|Ax_n - Ap\|^2\right\} \\
&\le \alpha_n\|u - p\|^2 + \|x_n - p\|^2 - \|z_n - x_n\|^2 \\
&\quad - 2\lambda_n\langle z_n - x_n, Ax_n - Ap\rangle - \lambda_n^2\|Ax_n - Ap\|^2.
\end{aligned} \tag{3.17}$$

Therefore, we have

$$\begin{aligned}
\|z_n - x_n\|^2 \le \alpha_n\|u - p\|^2 + \|x_n - p\|^2 - \|x_{n+1} - p\|^2 \\
- 2\lambda_n\langle z_n - x_n, Ax_n - Ap\rangle - \lambda_n^2\|Ax_n - Ap\|^2.
\end{aligned} \tag{3.18}$$

Replacing n by n_k, we have

$$\begin{aligned}
\|z_{n_k} - x_{n_k}\|^2 \le \alpha_{n_k}\|u - p\|^2 + \|x_{n_k} - p\|^2 - \|x_{n_k+1} - p\|^2 \\
- 2\lambda_{n_k}\langle z_{n_k} - x_{n_k}, Ax_{n_k} - Ap\rangle - \lambda_{n_k}^2\|Ax_{n_k} - Ap\|^2.
\end{aligned} \tag{3.19}$$

Since $\lim_{k\to\infty}\|x_{n_k} - p\|$ exists, from (3.14) and $\lim_{n\to\infty}\alpha_n = 0$, we obtain

$$\lim_{n\to\infty}\|z_{n_k} - x_{n_k}\| = 0. \tag{3.20}$$

Since A is Lipschitz continuous, we also obtain

$$\lim_{n\to\infty}\|Az_{n_k} - Ax_{n_k}\| = 0. \tag{3.21}$$

By the definition of J_{λ_n} and (3.1), we have that

$$\begin{aligned} z_n &= (I - \lambda_n B)^{-1}(I - \lambda_n A)x_n \\ &\Longleftrightarrow (I - \lambda_n A)x_n \in (I - \lambda_n B)z_n = z_n + \lambda_n B z_n \\ &\Longleftrightarrow x_n - z_n - \lambda_n A x_n \in \lambda_n B z_n \\ &\Longleftrightarrow \frac{1}{\lambda_n}(x_n - z_n - \lambda_n A x_n) \in B z_n. \end{aligned} \tag{3.22}$$

Since B is monotone, so for $(e, f) \in B$, we have that

$$\left\langle z_n - e, \frac{1}{\lambda_n}(x_n - z_n - \lambda_n A x_n) - f \right\rangle \geq 0, \tag{3.23}$$

and hence

$$\langle z_n - e, x_n - z_n - \lambda_n(Ax_n + f)\rangle \geq 0. \tag{3.24}$$

Replacing n by n_{k_i}, we have that

$$\left\langle z_{n_{k_i}} - e, x_{n_{k_i}} - z_{n_{k_i}} - \lambda_{n_{k_i}}\left(Ax_{n_{k_i}} + f\right)\right\rangle \geq 0. \tag{3.25}$$

Since A is an α-inverse strongly monotone, we have

$$\left\langle x_{n_{k_i}} - w, Ax_{n_{k_i}} - Aw \right\rangle \geq \alpha \left\| Ax_{n_{k_i}} - Aw \right\|^2. \tag{3.26}$$

This means that $Ax_{n_{k_i}} \to Aw$ as $i \to \infty$. From (3.20) and $x_{n_{k_i}} \rightharpoonup w$, we get that $z_{n_{k_i}} \rightharpoonup w$, together with (3.25), we have that

$$\langle w - e, -Aw - f\rangle \geq 0. \tag{3.27}$$

Since B is maximal monotone, so $(-Aw) \in Bw$. That is, $w \in (A + B)^{-1}(0)$.

Now, we prove that $x_n \to Pu$ as $n \to \infty$. Without loss of generality, we may assume that there exists a subsequence $\{x_{n_{k_i}+1}\}$ of $\{x_{n+1}\}$ such that

$$\limsup_{n\to\infty}\langle u - Pu, x_{n+1} - Pu\rangle = \lim_{i\to\infty}\left\langle u - Pu, x_{n_{k_i}+1} - Pu\right\rangle. \tag{3.28}$$

Since P is the metric projection of H onto F and $x_{n_{k_i}+1} \rightharpoonup w \in F$, we have

$$\lim_{i\to\infty}\left\langle u - Pu, x_{n_{k_i}+1} - Pu\right\rangle = \langle u - Pu, w - Pu\rangle \leq 0. \tag{3.29}$$

This implies that

$$\lim_{n \to \infty} \langle u - Pu, x_{n+1} - Pu \rangle \leq 0. \tag{3.30}$$

From (2.1), (3.1), and (3.3), we have

$$\begin{aligned} \|x_{n+1} - Pu\|^2 &= \|(1 - \alpha_n)(y_n - Pu) + \alpha_n(u - Pu)\|^2 \\ &\leq (1 - \alpha_n)^2 \|y_n - Pu\|^2 + 2\alpha_n \langle u - Pu, x_{n+1} - Pu \rangle \\ &\leq (1 - \alpha_n)\|x_n - Pu\|^2 + 2\alpha_n \langle u - Pu, x_{n+1} - Pu \rangle. \end{aligned} \tag{3.31}$$

From Lemma 2.5 and (3.30), we have

$$\lim_{n \to \infty} \|x_n - Pu\| = 0. \tag{3.32}$$

This means that $x_n \to Pu$ as $n \to \infty$. □

References

[1] F. E. Browder, "Convergence theorems for sequences of nonlinear operators in Banach spaces," *Mathematische Zeitschrift*, vol. 100, pp. 201–225, 1967.
[2] K. Goebel and W. A. Kirk, *Topics in Metric Fixed Point Theory*, vol. 28 of *Cambridge Studies in Advanced Mathematics*, Cambridge University Press, Cambridge, UK, 1990.
[3] F. Kohsaka and W. Takahashi, "Fixed point theorems for a class of nonlinear mappings related to maximal monotone operators in Banach spaces," *Archiv der Mathematik*, vol. 91, no. 2, pp. 166–177, 2008.
[4] S. Iemoto and W. Takahashi, "Approximating common fixed points of nonexpansive mappings and nonspreading mappings in a Hilbert space," *Nonlinear Analysis. Theory, Methods & Applications A*, vol. 71, no. 12, pp. e2082–e2089, 2009.
[5] T. Igarashi, W. Takahashi, and K. Tanaka, "Weak convergence theorems for nonspreading mappings and equilibrium problems," in *Nonlinear Analysis and Optimization*, S. Akashi, W. Takahashi, and T. Tanaka, Eds., pp. 75–85, Yokohama Publishers, Yokohama, Japan, 2009.
[6] W. Takahashi, "Fixed point theorems for new nonlinear mappings in a Hilbert space," *Journal of Nonlinear and Convex Analysis*, vol. 11, no. 1, pp. 79–88, 2010.
[7] W. R. Mann, "Mean value methods in iteration," *Proceedings of the American Mathematical Society*, vol. 4, pp. 506–510, 1953.
[8] A. Genel and J. Lindenstrauss, "An example concerning fixed points," *Israel Journal of Mathematics*, vol. 22, no. 1, pp. 81–86, 1975.
[9] B. Halpren, "Fixed points of nonexpansive maps," *Bulletin of the American Mathematical Society*, vol. 73, pp. 957–961, 1967.
[10] R. Wittmann, "Approximation of fixed points of nonexpansive mappings," *Archiv der Mathematik*, vol. 58, no. 5, pp. 486–491, 1992.
[11] J. B. Baillon, "Un theoreme de type ergodique pour les contractions non lineaires dans un espace de Hilbert," *Comptes Rendus De L'Acadḿie Des Sciences A-B*, vol. 280, pp. A1511–A1514, 1975.
[12] S. Takahashi, W. Takahashi, and M. Toyoda, "Strong convergence theorems for maximal monotone operators with nonlinear mappings in Hilbert spaces," *Journal of Optimization Theory and Applications*, vol. 147, no. 1, pp. 27–41, 2010.
[13] H. Manaka and W. Takahashi, "Weak convergence theorems for maximal monotone operators with nonspreading mappings in a Hilbert space," *Cubo*, vol. 13, no. 1, pp. 11–24, 2011.

[14] W. Takahashi, *Introduction to Nonlinear and Convex Analysis*, Yokohama Publishers, Yokohama, 2009.
[15] J. Schu, "Weak and strong convergence to fixed points of asymptotically nonexpansive mappings," *Bulletin of the Australian Mathematical Society*, vol. 43, no. 1, pp. 153–159, 1991.
[16] L. S. Liu, "Ishikawa and Mann iterative process with errors for nonlinear strongly accretive mappings in Banach spaces," *Journal of Mathematical Analysis and Applications*, vol. 194, no. 1, pp. 114–125, 1995.
[17] H.-K. Xu, "Iterative algorithms for nonlinear operators," *Journal of the London Mathematical Society*, vol. 66, no. 1, pp. 240–256, 2002.

5

Singular Initial Value Problem for a System of Integro-Differential Equations

Zdeněk Šmarda[1] and Yasir Khan[2]

[1] Department of Mathematics, Brno University of Technology, 61600 Brno, Czech Republic
[2] Department of Mathematics, Zhejiang University, Hangzhou 310027, China

Correspondence should be addressed to Zdeněk Šmarda, smarda@feec.vutbr.cz

Academic Editor: Juntao Sun

Analytical properties like existence, uniqueness, and asymptotic behavior of solutions are studied for the following singular initial value problem: $g_i(t)y_i'(t) = a_i y_i(t)(1 + f_i(t, \mathbf{y}(t), \int_{0^+}^t K_i(t, s, \mathbf{y}(t), \mathbf{y}(s))ds))$, $y_i(0^+) = 0$, $t \in (0, t_0]$, where $\mathbf{y} = (y_1, \ldots, y_n)$, $a_i > 0$, $i = 1, \ldots, n$ are constants and $t_0 > 0$. An approach which combines topological method of T. Ważewski and Schauder's fixed point theorem is used. Particular attention is paid to construction of asymptotic expansions of solutions for certain classes of systems of integrodifferential equations in a right-hand neighbourhood of a singular point.

1. Introduction and Preliminaries

Singular initial value problem for ordinary differential and integro-differential equations is fairly well studied (see, e.g., [1–16]), but the asymptotic properties of the solutions of such equations are only partially understood. Although the singular initial value problems were widely considered using various methods (see, e.g., [1–13, 16]), our approach to this problem is essentially different from others known in the literature. In particular, we use a combination of the topological method of T. Ważewski [8] and Schauder's fixed point theorem [11]. Our technique leads to the existence and uniqueness of solutions with asymptotic estimates in the right-hand neighbourhood of a singular point. Asymptotic expansions of solutions are constructed for certain classes of systems of integrodifferential equations as well.

Consider the following problem:

$$g_i(t)y_i'(t) = a_i y_i(t)\left(1 + f_i\left(t, \mathbf{y}(t), \int_{0^+}^t K_i(t, s, \mathbf{y}(t), \mathbf{y}(s))ds\right)\right), \tag{1.1}$$

$$y_i(0^+) = 0, \quad t \in (0, t_0], \tag{1.2}$$

where $\mathbf{y} = (y_1, \ldots, y_n)$, $a_i > 0$ are constants, $f_i \in C^0(J \times \mathbb{R}^n \times \mathbb{R}, \mathbb{R})$, $K_i \in C^0(J \times J \times \mathbb{R}^n \times \mathbb{R}^n, \mathbb{R})$, $J = (0, t_0]$, $t_0 > 0$, $i = 1, \ldots, n$.

Denote

(i) $f(t) = O(g(t))$ as $t \to 0^+$ if there is a right-hand neighbourhood $\mathcal{U}(0)$ and a constant $K > 0$ such that $(f(t)/g(t)) \leq K$ for $t \in \mathcal{U}(0)$.

(ii) $f(t) = o(g(t))$ as $t \to 0^+$ if there is valid $\lim_{t \to 0^+} f(t)/g(t) = 0$.

(iii) $f(t) \sim g(t)$ as $t \to 0^+$ if there is valid $\lim_{t \to 0^+} f(t)/g(t) = 1$.

Definition 1.1. The sequence of functions $(\phi_n(t))$ is called an asymptotic sequence as $t \to 0^+$ if

$$\phi_{n+1}(t) = o(\phi_n(t)) \quad \text{as } t \to 0^+ \tag{1.3}$$

for all n.

Definition 1.2. The series $\sum c_n \phi_n(t)$, $c_n \in \mathbb{R}$, is called an asymptotic expansion of the function $f(t)$ up to Nth term as $t \to 0^+$ if

(a) $(\phi_n(t))$ is an asymptotic sequence,

(b)

$$\left[f(t) - \sum_{n=1}^{N} c_n \phi_n(t) \right] = o(\phi_N(t)), \quad \text{as } t \to 0^+. \tag{1.4}$$

The functions g_i, f_i, and K_i will be assumed to satisfy the following:

(i) $g_i(t) \in C^1(J)$, $g_i(t) > 0$, $g_i(0^+) = 0$, $g_i'(t) \sim \varphi_i(t) g_i^{\lambda_i}(t)$ as $t \to 0^+$, $\lambda_i > 0$, $\varphi_i(t) g_i^{\tau}(t) = o(1)$ as $t \to 0^+$ for each $\tau > 0$, $i = 1, \ldots, n$,

(ii) $|f_i(t, u, v)| \leq |u| + |v|$, $|\int_{0^+}^{t} K_i(t, s, \mathbf{y}(t), \mathbf{y}(s)) ds| \leq r_i(t)|\mathbf{y}|$, $0 < r_i(t) \in C(J)$, $r_i(t) = \varphi_i(t, C_i) o(1)$ as $t \to 0^+$ where $\varphi_i(t, C_i) = C_i \exp(\int_{t_0}^{t} (a_i / g_i(s)) ds)$ is the general solution of the equation $g_i(t) y_i'(t) = a_i y_i(t)$.

In the text, we will apply topological method of Ważewski and Schauder's theorem. Therefore we give a short summary of them.

Let $f(t, \mathbf{y})$ be a continuous function defined on an open $(t, \mathbf{y})$ set $\Omega \subset \mathbb{R} \times \mathbb{R}^n$, Ω^0 an open set of Ω, $\partial \Omega^0$ the boundary of Ω^0, and $\overline{\Omega}^0$ the closure of Ω^0. Consider the following system of ordinary differential equations:

$$\mathbf{y}' = f(t, \mathbf{y}). \tag{1.5}$$

Definition 1.3 (see [17]). The point $(t_0, \mathbf{y}_0) \in \Omega \cap \partial \Omega^0$ is called an egress (or an ingress point) of Ω^0 with respect to system (1.5) if for every fixed solution of the problem $\mathbf{y}(t_0) = \mathbf{y}_0$, there exists an $\epsilon > 0$ such that $(t, \mathbf{y}(t)) \in \Omega^0$ for $t_0 - \epsilon \leq t < t_0$ ($t_0 < t \leq t_0 + \epsilon$). An egress point (ingress point) $(t_0, \mathbf{y}_0)$ of Ω^0 is called a strict egress point (strict ingress point) of Ω^0 if $(t, \mathbf{y}(t)) \notin \overline{\Omega}^0$ on interval $t_0 < t \leq t_0 + \epsilon_1$ ($t_0 - \epsilon_1 \leq t < t_0$) for an ϵ_1.

Definition 1.4 (see [18]). An open subset Ω^0 of the set Ω is called an (u, v) subset of Ω with respect to system (1.5) if the following conditions are satisfied.

(1) There exist functions $u_i(t, \mathbf{y}) \in C^1(\Omega, \mathbb{R})$, $i = 1, \ldots, m$ and $v_j(t, \mathbf{y}) \in C[\Omega, \mathbb{R}]$ $j = 1, \ldots, n, m + n > 0$ such that

$$\Omega_0 = \{(t, \mathbf{y}) \in \Omega : u_i(t, \mathbf{y}) < 0, v_j(t, \mathbf{y}) < 0\ \forall i, j\}. \tag{1.6}$$

(2) $\dot{u}_\alpha(t, \mathbf{y}) < 0$ holds for the derivatives of the functions $u_\alpha(t, \mathbf{y})$, $\alpha = 1, \ldots, m$ along trajectories of system (1.5) on the set

$$U_\alpha = \{(t, \mathbf{y}) \in \Omega : u_\alpha(t, \mathbf{y}) = 0, u_i(t, \mathbf{y}) \leq 0, v_j(t, \mathbf{y}) \leq 0, \forall j \text{ and } i \neq \alpha\}. \tag{1.7}$$

(3) $\dot{v}_\beta(t, \mathbf{y}) > 0$ holds for the derivatives of the functions $v_\beta(t, \mathbf{y})$, $\beta = 1, \ldots, n$ along trajectories of system (1.5) on the set

$$V_\beta = \{(t, \mathbf{y}) \in \Omega : u_\beta(t, \mathbf{y}) = 0, u_i(t, \mathbf{y}) \leq 0, v_j(t, y) \leq 0, \forall i \text{ and} j \neq \beta\}. \tag{1.8}$$

The set of all points of egress (strict egress) is denoted by Ω_e^0 (Ω_{se}^0).

Lemma 1.5 (see [18]). *Let the set Ω_0 be a (u, v) subset of the set Ω with respect to system* (1.5). *Then*

$$\Omega_{\text{se}}^0 = \Omega_e^0 = \bigcup_{\alpha=1}^{m} U_\alpha \setminus \bigcup_{\beta=1}^{n} V_\beta. \tag{1.9}$$

Definition 1.6 (see [18]). Let X be a topological space and $B \subset X$.

Let $A \subset B$. A function $r \in C(B, A)$ such that $r(a) = a$ for all $a \in A$ is a retraction from B to A in X.

The set $A \subset B$ is a retract of B in X if there exists a retraction from B to A in X.

Theorem 1.7 (Ważewski's theorem [18]). *Let Ω^0 be some (u, v) subset of Ω with respect to system* (1.5). *Let S be a nonempty compact subset of $\Omega^0 \cup \Omega_e^0$ such that the set $S \cap \Omega_e^0$ is not a retract of S but is a retract Ω_e^0. Then there is at least one point $(t_0, \mathbf{y}_0) \in S \cap \Omega_0$ such that the graph of a solution $\mathbf{y}(t)$ of the Cauchy problem $\mathbf{y}(t_0) = \mathbf{y}_0$ for* (1.5) *lies on its right-hand maximal interval of existence.*

Theorem 1.8 (Schauder's theorem [19]). *Let E be a Banach space and S its nonempty convex and closed subset. If P is a continuous mapping of S into itself and PS is relatively compact then the mapping P has at least one fixed point.*

2. Main Results

Theorem 2.1. *Let assumptions (i) and (ii) hold, then for each* $C_i \neq 0$ *there is one solution* $\mathbf{y}(t,\mathbf{C}) = (y_1(t,C_1), y_2(t,C_2), \ldots, y_n(t,C_n))$, $\mathbf{C} = (C_1,\ldots,C_n)$ *of initial problem* (1.1) *and* (1.2) *such that*

$$\left|y_i^{(j)}(t,C_i) - \varphi_i^{(j)}(t,C_i)\right| \leq \delta\left(\varphi_i^2(t,C_i)\right)^{(j)}, \quad j = 0,1, \tag{2.1}$$

for $t \in (0,t^{\Delta}]$, *where* $0 < t^{\Delta} \leq t_0$, $\delta > 1$ *is a constant, and* t^{Δ} *depends on* δ, C_i, $i = 1,\ldots,n$.

Proof. (1) Denote E the Banach space of vector-valued continuous functions $\mathbf{h}(t)$ on the interval $[0,t_0]$ with the norm

$$\|\mathbf{h}(t)\| = \max_{t\in[0,t_0]} |h_i(t)|, \quad i = 1,\ldots,n. \tag{2.2}$$

The subset S of Banach space E will be the set of all functions $h(t)$ from E satisfying the inequality

$$\left|h_i(t) - \varphi_i(t,C_i)\right| \leq \delta\varphi_i^2(t,C_i). \tag{2.3}$$

The set S is nonempty, convex, and closed.

(2) Now we will construct the mapping P. Let $\mathbf{h}_0(t) \in S$ be an arbitrary function. Substituting $\mathbf{h}_0(t)$, $\mathbf{h}_0(s)$ instead of $\mathbf{y}(t)$, $\mathbf{y}(s)$ into (1.1), we obtain the following differential equation:

$$g_i(t)y_i'(t) = a_i y_i(t)\left(1 + f_i\left(t,\mathbf{y}(t),\int_{0^+}^t K_i(t,s,\mathbf{h}_0(t),\mathbf{h}_0(s))ds\right)\right), \quad i = 1,\ldots,n. \tag{2.4}$$

Put

$$y_i(t) = \varphi_i(t,C_i) + \varphi_i^{(1-\mu)}(t,C_i)Y_{0i}(t), \tag{2.5}$$

$$y_i'(t) = \varphi_i'(t,C) + \frac{1}{g_i(t)}\varphi_i^{(1-\mu)}(t,C_i)Y_{1i}(t), \tag{2.6}$$

where $0 < \mu < 1$ is a constant and new functions $Y_{0i}(t)$, $Y_{1i}(t)$ satisfy the differential equations as

$$g_i(t)Y_{0i}'(t) = (\mu - 1)a_i Y_{0i}(t) + Y_{1i}(t), \quad i = 1,\ldots,n. \tag{2.7}$$

From (2.3), it follows

$$h_{0i}(t) = \varphi_i(t,C_i) + H_{0i}(t), \quad |H_{0i}(t)| \leq \delta\varphi_i^2(t,C_i). \tag{2.8}$$

Substituting (2.5), (2.6), and (2.8) into (2.4), we get

$$Y_{1i}(t) = a_i Y_{0i}(t) + \left(a_i \varphi_i^{\mu}(t, C_i) + a_i Y_{0i}(t)\right) \times f_i\Bigg(t, \varphi_1(t, C_1) + \varphi_1^{(1-\mu)}(t, C_1) Y_{01}(t), \dots, \varphi_n(t, C_n) + \varphi_n^{(1-\mu)}(t, C_n) Y_{0n}(t)\ , \int_{0^+}^{t} K_i(t, s, \varphi_1(t, C_1) + H_{01}(t), \dots, \varphi_n(t, C_n) + H_{0n}(t), \varphi_1(s, C_1) + H_{01}(s), \dots \varphi_n(s, C_n) + H_{0n}(s)) ds\Bigg). \tag{2.9}$$

Substituting (2.9) into (2.7), we get

$$g_i(t) Y'_{0i}(t) = \mu a_i Y_{0i}(t) + \left(a_i \varphi_i^{\mu}(t, C_i) + a_i Y_{0i}(t)\right) \times f_i\Bigg(t, \varphi_1(t, C_1) + \varphi_1^{(1-\mu)}(t, C_1) Y_{01}(t), \dots, \varphi_n(t, C_n) + \varphi_n^{(1-\mu)}(t, C_n) Y_{0n}(t), \int_{0^+}^{t} K_i(t, s, \varphi_1(t, C_1) + H_{01}(t), \dots, \varphi_n(t, C_n) + H_{0n}(t), \varphi_1(s, C_1) + H_{01}(s), \dots \varphi_n(s, C_n) + H_{0n}(s)) ds\Bigg). \tag{2.10}$$

In view of (2.5) and (2.6), it is obvious that a solution of (2.10) determines a solution of (2.4).

Now we use Ważewski's topological method. Consider an open set $\Omega \subset \mathbb{R}^+ \times \mathbb{R}^n$. Denote $\mathbf{Y}_0 = (Y_{01}, \dots, Y_{0n})$. Define an open subset $\Omega_0 \subset \Omega$ as follows:

$$\begin{aligned} \Omega_0 &= \{(t, \mathbf{Y}_0) : u_i(t, \mathbf{Y}_0) < 0, v(t, \mathbf{Y}_0) < 0, i = 1, \dots, n\}, \\ U_\alpha &= \{(t, \mathbf{Y}_0) : u_\alpha(t, \mathbf{Y}_0) = 0, u_i(t, \mathbf{Y}_0) \le 0, v(t, \mathbf{Y}_0) \le 0, i = 1, \dots, n, i \neq \alpha\}, \\ V_\beta = V &= \{(t, \mathbf{Y}_0) : v(t, \mathbf{Y}_0) = 0, u_j(t, \mathbf{Y}_0) \le 0, i = 1, \dots, n\}, \end{aligned} \tag{2.11}$$

where

$$u_i(t, \mathbf{Y}_0) = Y_{0i}^2 - \left(\delta \varphi_i^{(1+\mu)}(t, C_i)\right)^2, \quad v(t, \mathbf{Y}_0) = t - t_0, \quad i = 1, \dots, n. \tag{2.12}$$

Calculating the derivatives $\dot{u}_\alpha(t,\mathbf{Y}_0)$, $\dot{v}(t,\mathbf{Y}_0)$ along the trajectories of (2.10) on the set U_α, V, $\alpha = 1,\ldots,n$ we obtain

$$\begin{aligned}\dot{u}_\alpha(t,\mathbf{Y}_0) = \frac{2a_\alpha}{g_\alpha(t)}\Bigg[&\mu Y_{0\alpha}^2(t) + \left(Y_{0\alpha}(t)\varphi_\alpha^\mu(t,C_\alpha) + Y_{0\alpha}^2(t)\right)\\ &\times f_\alpha\Bigg(t,\varphi_1(t,C_1) + \varphi_1^{(1-\mu)}(t,C_1)Y_{01}(t),\ldots,\varphi_n(t,C_n) + \varphi_n^{(1-\mu)}(t,C_n)Y_{0n}(t),\\ &\int_{0^+}^t K_i(t,s,\varphi_1(t,C_1) + H_{01}(t),\ldots,\varphi_n(t,C_n)\\ &+H_{0n}(t),\varphi_1(s,C_1) + H_{01}(s),\ldots\varphi_n(s,C_n) + H_{0n}(s))ds\Bigg).\\ &-\delta^2(1+\mu)\varphi_\alpha^{2(1+\mu)}(t,C_\alpha)\Bigg].\end{aligned} \tag{2.13}$$

Since

$$\begin{aligned}&\lim_{t\to+0}\varphi_i(t)g_i^\tau(t) = 0 \quad \text{for any } \tau > 0,\ i = 1,\ldots,n\\ &g_i'(t) \sim \varphi_i(t)g_i^{\lambda_i}(t) \quad \text{as } t\to 0^+,\ \lambda_i > 0,\ i = 1,\ldots,n,\end{aligned} \tag{2.14}$$

then there exists a positive constant M_i such that

$$g_i'(t) < M_i, \quad t\in(0,t_0],\ i = 1,\ldots,n. \tag{2.15}$$

Consequently,

$$\int_{t_0}^t \frac{ds}{g_i(s)} < \frac{1}{M_i}\int_{t_0}^t \frac{g_i'(s)dt}{g_i(s)} = \frac{1}{M_i}\ln\frac{g_i(t)}{g_i(t_0)} \longrightarrow -\infty \quad \text{as } t\longrightarrow 0^+,\ i = 1,\ldots,n. \tag{2.16}$$

From here $\lim_{t\to 0^+}\varphi_i(t,C_i) = 0$ and by L'Hospital's rule $\varphi_i^\tau(t,C_i)g_i^\sigma(t) = o(1)$, for $t \to 0^+$, $i = 1,\ldots,n,\sigma$ is an arbitrary real number. These both identities imply that the powers of $\varphi_i(t,C_i)$ affect the convergence to zero of the terms in (2.13), in a decisive way.

Using the assumptions of Theorem 2.1 and the definition of $Y_0(t)$, $\varphi_i(t,C_i)$, $i = 1,\ldots,n$, we get that the first term $\mu Y_{0\alpha}^2(t,C_\alpha)$ in (2.13) has the following form:

$$\mu Y_{0\alpha}^2(t) = \mu\delta^2\varphi_\alpha^{2(1+\mu)}(t,C_\alpha), \tag{2.17}$$

and the second term

$$\begin{aligned}
&\left(Y_{0\alpha}(t)\varphi_\alpha^\mu(t,C_\alpha)+Y_{0\alpha}^2(t)\right)\\
&\quad\times f_\alpha\Big(t,\varphi_1(t,C_1)+\varphi_1^{(1-\mu)}(t,C_1)Y_{01}(t),\dots,\varphi_n(t,C_n)\\
&\qquad+\varphi_n^{(1-\mu)}(t,C_n)Y_{0n}(t),\int_{0^+}^t K_\alpha(t,s,\varphi_1(t,C_1)+H_{01}(t),\dots,\varphi_n(t,C_n)\\
&\qquad\qquad+H_{0n}(t),\varphi_1(s,C_1)+H_{01}(s),\dots\varphi_n(s,C_n)\\
&\qquad\qquad+H_{0n}(s))ds\Big).
\end{aligned} \tag{2.18}$$

is bounded by terms with exponents which are greater than $\varphi_\alpha^{2(1+\mu)}(t,C_\alpha)$, $\alpha=1,\dots,n$. From here, we obtain

$$\operatorname{sgn}\dot{u}_\alpha(t,\mathbf{Y}_0)=-\delta^2(1+\mu)\varphi_\alpha^{2(1+\mu)}(t,C_\alpha)=-1 \tag{2.19}$$

for sufficiently small t^*, depending on C_α $\alpha=1,\dots,n$, δ, $0<t^*\le t_0$.

It is obvious that $\operatorname{sgn}\dot{v}(t,\mathbf{Y}_0)=1$.

Change the orientation of the axis t into opposite. Then, with respect to the new system of coordinates, the set Ω_0 is the (u,v) subset with respect to system (2.10). By Ważewski's topological method, we state that there exists at least one integral curve of (2.10) lying in Ω_0 for $t\in(0,t^*)$. It is obvious that this assertion remains true for an arbitrary function $\mathbf{h}_0(t)\in S$.

Now we prove the uniqueness of a solution of (2.10). Let $\overline{\mathbf{Y}}_0(t)=(\overline{Y}_{01}(t),\dots,\overline{Y}_{0n}(t))$ be also the solution of (2.10). Putting

$$Z_{0i}=Y_{0i}-\overline{Y}_{0i},\quad i=1,\dots,n \tag{2.20}$$

and substituting into (2.10), we obtain

$$\begin{aligned}
g_i(t)Z_{0i}'(t)&=\mu a_iY_{0i}(t)+\left(a_i\varphi_i^\mu(t,C_i)+a_iZ_{0i}(t)\right)\\
&\quad\times f_i\Big(t,\varphi_1(t,C_1)+\varphi_1^{(1-\mu)}(t,C_1)\left(Z_{0i}(t)+\overline{Y}_{01}(t)\right),\dots,\varphi_n(t,C_n)\\
&\qquad+\varphi_n^{(1-\mu)}(t,C_n)\left(Z_{0n}(t)+\overline{Y}_{0n}(t)\right),\\
&\qquad\int_{0^+}^t K_i(t,s,\varphi_1(t,C_1)+H_{01}(t),\dots,\varphi_n(t,C_n)\\
&\qquad\qquad+H_{0n}(t),\varphi_1(s,C_1)+H_{01}(s),\dots\varphi_n(s,C_n)+H_{0n}(s))ds\Big).
\end{aligned} \tag{2.21}$$

Define

$$\begin{aligned}\Omega_1(\delta) &= \{(t,\mathbf{Z}_0): 0<t<t^*, u_{1i}(t,\mathbf{Z}_0)<0, v_1(t,\mathbf{Z}_0)<0, 0<t<t^*, i=1,\ldots,n\}\\ U_{1\alpha} &= \{(t,\mathbf{Z}_0): u_{1\alpha}(t,\mathbf{Z}_0)=0, u_{1i}(t,\mathbf{Z}_0)\le 0, v_1(t,\mathbf{Z}_0)\le 0, i=1,\ldots,n, i\neq\alpha\},\\ V_{1\beta} &= V = \{(t,\mathbf{Z}_0): v_1(t,\mathbf{Z}_0)=0, u_j(t,\mathbf{Z}_0)\le 0, i=1,\ldots,n\},\end{aligned} \tag{2.22}$$

where

$$u_{1i}(t,\mathbf{Z}_0) = Z_{0i}^2 - \left(\delta\varphi_i^{(1+\mu-\gamma)}\right)^2, \quad 0<\gamma<\mu,\ v_1(t,\mathbf{Z}_0) = t-t^*. \tag{2.23}$$

Using the same method as above, we have

$$\operatorname{sgn}\dot{u}_{1i}(t,\mathbf{Z}_0) = -1, \qquad \operatorname{sgn}\dot{v}_1(t,\mathbf{Z}_0) = 1, \quad i=1,\ldots,n \tag{2.24}$$

for sufficiently small $t^\diamond$, $0<t^\diamond\le t^*$. It is obvious that $\Omega_0\subset\Omega_1(\delta)$ for $t\in(0,t^\diamond)$. Let $\overline{\mathbf{Z}}_0(t) = (\overline{Z}_{01}(t),\ldots,\overline{Z}_{0n}(t))$ be any nonzero solution of (2.10) such that $(t_1,\overline{\mathbf{Z}_0}(t_1))\in\Omega_1$ for $0<t_1<t^\diamond$. Let $\overline{\delta}\in(0,\delta)$ be such a constant that $(t_1,\overline{\mathbf{Z}_0}(t_1))\in\partial\Omega_1(\overline{\delta})$. If the curve $\overline{\mathbf{Z}_0}(t)$ lay in $\Omega_1(\overline{\delta})$ for $0<t<t_1$, then $(t_1,\overline{\mathbf{Z}_0}(t_1))$ would have to be a strict egress point of $\partial\Omega_1(\overline{\delta})$ with respect to the original system of coordinates. This contradicts the relation (2.24). Therefore there exists only the trivial solution $\mathbf{Z}_0(t)\equiv 0$ of (2.21), so $\mathbf{Y}_0 = \overline{\mathbf{Y}_0}(t)$ is the unique solution of (2.10).

From (2.5) we obtain

$$\left|y_i(t,C_i)-\varphi_i(t,C_i)\right| \le \delta\varphi_i^2(t,C_i), \quad i=1,\ldots,n, \tag{2.25}$$

where $(y_1(t,C_1),\ldots,y_n(t,C_n))$ is the solution of (2.4) for $t\in(0,t^\diamond]$. Similarly, from (2.6) and (2.9), we have

$$\begin{aligned}\left|y_i'(t,C_i)-\varphi_i'(t,C_i)\right| &= \left|\frac{1}{g_i(t)}\varphi_i^{(1-\mu)}(t,C_i)Y_{1i}(t)\right|\\ &\le \left|\frac{1}{g_i(t)}\varphi_i^{(1-\mu)}(t,C_i)2a_i\delta\varphi_i^{(1-\mu)}(t,C_i)\right| = \delta\left(\varphi_i^2(t,C_i)\right)'.\end{aligned} \tag{2.26}$$

It is obvious (after a continuous extension of $\mathbf{y}(t,\mathbf{C})$ for $t=0$, $\mathbf{y}(0^+)=0$) that $P:\mathbf{h}_0\to\mathbf{y}$ maps S into itself and $PS\subset S$.

(3) We will prove that PS is relatively compact and P is a continuous mapping.

It is easy to see, by (2.25) and (2.26), that PS is the set of uniformly bounded and equicontinuous functions for $t\in[0,t^\diamond]$. By Ascoli's theorem, PS is relatively compact.

Let $\{\mathbf{h}_k(t)\}$ be an arbitrary sequence vector-valued functions in S such that

$$\|\mathbf{h}_k(t)-\mathbf{h}_0(t)\| = \epsilon_k, \qquad \lim_{k\to\infty}\epsilon_k = 0, \quad \mathbf{h}_0(t)\in S. \tag{2.27}$$

The solution $\overline{\mathbf{Y}}_k(t) = (\overline{Y}_{k1}, \dots, \overline{Y}_{kn})$ of the following equation:

$$
\begin{aligned}
g_i(t)Y_{0i}'(t) = \mu a_i Y_{0i}(t) + \left(a_i \varphi_i^{\mu}(t, C_i) + a_i Y_{0i}(t)\right) \\
\times f_i\Bigg(t, \varphi_1(t, C_1) + \varphi_1^{(1-\mu)}(t, C_1)Y_{01}(t), \dots, \varphi_n(t, C_n) + \varphi_n^{(1-\mu)}(t, C_n)Y_{0n}(t), \\
\int_{0^+}^{t} K_i(t, s, \varphi_1(t, C_1) + H_{01}(t), \dots, \varphi_n(t, C_n) \\
+H_{0n}(t), \varphi_1(s, C_1) + H_{k1}(s), \dots \varphi_n(s, C_n) + H_{kn}(s))ds\Bigg).
\end{aligned}
\tag{2.28}
$$

corresponds to the function $\mathbf{h}_k(t)$ and $\overline{\mathbf{Y}_k}(t) \in \Omega_0$ for $t \in (0, t^{\diamond})$. Similarly, the solution $\overline{\mathbf{Y}_0}(t)$ of (2.10) corresponds to the function $h_0(t)$. We will show that $|\overline{\mathbf{Y}_k}(t) - \overline{\mathbf{Y}_0}(t)| \to 0$ uniformly on $[0, t^{\Delta}]$, where $0 < t^{\Delta} \le t^{\diamond}$, t^{Δ} is a sufficiently small constant which will be specified later. Consider the following region:

$$
\Omega_{0k} = \left\{(t, \mathbf{Y}_0) : 0 < t < t^{\diamond}, u_{0k_i}(t, \mathbf{Y}_0) < 0, v_0(t, \mathbf{Y}_0) < 0, i = 1, \dots, n\right\}, \tag{2.29}
$$

where

$$
\begin{gathered}
u_{0k_i}(t, \mathbf{Y}_0) = \left(Y_{0i}(t) - \overline{Y}_{0i}(t)\right)^2 - \left(\epsilon_k \varphi_i^{(1+\mu-\nu)}(t, C_i)\right)^2, \quad 0 < \nu < \alpha,\ i = 1, \dots, n,\ k \ge 1, \\
v_0(t, \mathbf{Y}_0) = t - t^{\diamond}.
\end{gathered}
\tag{2.30}
$$

There exists sufficiently small constant $t^{\Delta} \le t^{\diamond}$ such that $\Omega_0 \subset \Omega_{0k}$ for any $k, t \in (0, t^{\Delta})$. Investigate the behaviour of integral curves of (2.28) with respect to the boundary $\partial\Omega_{0k}$, $t \in (0, t^{\Delta}]$. Using the same method as above, we obtain the following trajectory derivatives:

$$
\operatorname{sgn} \dot{u}_{0k}(t, \mathbf{Y}_0) = -1, \qquad \operatorname{sgn} \dot{v}_0(t, \mathbf{Y}_0) = 1 \tag{2.31}
$$

for $t \in (0, t^{\Delta}]$ and any k. By Ważewski's topological method, there exists at least one solution $\overline{\mathbf{Y}}_k(t)$ lying in Ω_{0k}, $0 < t < t^{\Delta}$. Hence, it follows that

$$
\left|\overline{Y}_{ki}(t) - \overline{Y}_{0i}(t)\right| \le \epsilon_k \varphi_i^{1+\mu-\nu} \le N_i \epsilon_k, \tag{2.32}
$$

$N_i > 0$, $i = 1, \dots, n$ are constants depending on C_i, t^{Δ}. From (2.5), we obtain

$$
\left|y_{ki}(t) - y_{0i}(t)\right| = \varphi_i^{(1-\mu)}(t, C_i)\left|\overline{Y}_{ki}(t) - \overline{Y}_{0i}(t)\right| \le n_i \epsilon_k, \tag{2.33}
$$

where $n_i > 0$, $i = 1, \dots, n$ are constants depending on t^{Δ}, C_i, N_i. This estimate implies that P is continuous.

We have thus proved that the mapping P satisfies the assumptions of Schauder's fixed point theorem and hence there exists a function $\mathbf{h}(t) \in S$ with $\mathbf{h}(t) = P(\mathbf{h}(t))$. The proof of existence of a solution of (1.1) is complete.

Now we will prove the uniqueness of a solution of (1.1). Substituting (2.5) and (2.6) into (1.1), we get

$$\begin{aligned} Y_{1i}(t) &= a_i Y_{0i}(t) + \left(a_i \varphi_i^{\mu}(t, C_i) + a_i Y_{0i}(t)\right) \\ &\times f_i\Bigg(t, \varphi_1(t, C_1) + \varphi_1^{(1-\mu)}(t, C_1) Y_{01}(t), \ldots, \varphi_n(t, C_n) + \varphi_n^{(1-\mu)}(t, C_n) Y_{0n}(t), \\ &\int_{0^+}^{t} K_i\Big(t, s, \varphi_1(t, C_1) + \varphi_1^{(1-\mu)}(t, C_1) Y_{01}(t), \ldots, \varphi_n(t, C_n) \\ &+ \varphi_n^{(1-\mu)}(t, C_n) Y_{0n}(t), \varphi_1(s, C_1) + \varphi_1^{(1-\mu)}(s, C_1) Y_{01}(s), \ldots \varphi_n(s, C_n) \\ &+ \varphi_n^{(1-\mu)}(s, C_n) Y_{0n}(s)\Big) ds\Bigg). \end{aligned} \tag{2.34}$$

Equation (2.7) may be written in the following form:

$$\begin{aligned} g_i(t) Y_{0i}'(t) &= a_i Y_{0i}(t) + \left(a_i \varphi_i^{\mu}(t, C_i) + a_i Y_{0i}(t)\right) \\ &\times f_i\Bigg(t, \varphi_1(t, C_1) + \varphi_1^{(1-\mu)}(t, C_1) Y_{01}(t), \ldots, \varphi_n(t, C_n) + \varphi_n^{(1-\mu)}(t, C_n) Y_{0n}(t), \\ &\int_{0^+}^{t} K_i\Big(t, s, \varphi_1(t, C_1) + \varphi_1^{(1-\mu)}(t, C_1) Y_{01}(t), \ldots, \varphi_n(t, C_n) + \varphi_n^{(1-\mu)}(t, C_n) \\ &\times Y_{0n}(t), \varphi_1(s, C_1) + \varphi_1^{(1-\mu)}(s, C_1) Y_{01}(s), \ldots \varphi_n(s, C_n) \\ &+ \varphi_n^{(1-\mu)}(s, C_n) Y_{0n}(s)\Big) ds\Bigg). \end{aligned} \tag{2.35}$$

Now we know that there exists the solution $\mathbf{y}_0(t) = (y_{01}(t, C_1), \ldots, y_{0n}(t, C_n))$ of (1.1) satisfying (1.2) such that

$$y_{0i}(t, C_i) = \varphi_i(t, C_i) + \varphi_i^{(1-\mu)}(t, C_i) T_{0i}(t), \quad i = 1, \ldots, n, \tag{2.36}$$

where $\mathbf{T}_0(t) = (T_{01}(t), \ldots, T_{0n}(t))$ is the solution of (2.35).

Denote $W_{i0}(t) = Y_{0i}(t) - T_{0i}(t)$, $i = 1, \dots, n$. Substituting $W_{i0}(t)$ into (2.35), we obtain

$$
\begin{aligned}
g_i(t)W'_{0i}(t) = {} & a_i W_{0i}(t) + \left(a_i\varphi_i^{\mu}(t, C_i) + a_i W_{0i}(t)\right) \\
& \times \Bigg[f_i\Bigg(t, \varphi_1(t, C_1) + \varphi_1^{(1-\mu)}(t, C_1)(W_{01}(t) + T_{01}(t)), \dots, \varphi_n(t, C_n) \\
& \qquad + \varphi_n^{(1-\mu)}(t, C_n) \times (W_{0n}(t) + T_{0n}(t)), \\
& \qquad \int_{0^+}^{t} K_i\Big(t, s, \varphi_1(t, C_1) + \varphi_1^{(1-\mu)}(t, C_1)(W_{01}(t) + T_{01}(t)), \dots, \varphi_n(t, C_n) \\
& \qquad\quad + \varphi_n^{(1-\mu)}(t, C_n) \times (W_{0n}(t) + T_{0n}(t)), \varphi_1(s, C_1) + \varphi_1^{(1-\mu)}(s, C_1) \\
& \qquad\quad \times (W_{01}(s) + T_{01}(s)), \dots \varphi_n(s, C_n) \\
& \qquad\quad + \varphi_n^{(1-\mu)}(s, C_n)(W_{0n}(s) + T_{0n}(s))\Big) ds \Bigg) \\
& - f_i\Bigg(t, \varphi_1(t, C_1) + \varphi_1^{(1-\mu)}(t, C_1)T_{01}(t), \dots, \varphi_n(t, C_n) + \varphi_n^{(1-\mu)}(t, C_n), \\
& \qquad \int_{0^+}^{t} K_i\Big(t, s, \varphi_1(t, C_1) + \varphi_1^{(1-\mu)}(t, C_1)T_{01}(t), \dots, \varphi_n(t, C_n)T_{0n}(t) \\
& \qquad\quad + \varphi_n^{(1-\mu)}(t, C_n)T_{0n}(t), \varphi_1(s, C_1) + \varphi_1^{(1-\mu)}(s, C_1) \\
& \qquad\quad \times T_{01}(s), \dots \varphi_n(s, C_n) + \varphi_n^{(1-\mu)}(s, C_n)T_{0n}(s)\Big) ds \Bigg) \Bigg].
\end{aligned}
\tag{2.37}
$$

Let

$$
{}^1\Omega_0 = \left\{ (t, \mathbf{W}_0) : 0 < t < t^{\Delta}, u_{1i}(t, \mathbf{W}_0) < 0, v_1(t, \mathbf{W}_0) < 0 \right\}, \tag{2.38}
$$

where

$$
u_{1i}(t, \mathbf{W}_0) = W_{1i}^2 - \left(\delta\varphi_i^{(1+\mu-\rho)}(t, C_i)\right)^2, \quad 0 < \rho < \mu, \quad v_1(t, \mathbf{W}_0) = t - t^{\Delta}, \quad i = 1, \dots, n. \tag{2.39}
$$

If (2.37) had only the trivial solution lying in ${}^1\Omega_0$, then $\mathbf{Y}_0(t) = \mathbf{T}_0(t)$ would be only one solution of (2.37) and from here, by (2.35), $\mathbf{y}_0(t)$ would be only one solution of (1.1) satisfying (1.2) for $t \in (0, t^{\Delta}]$.

We will suppose that there exists nontrivial solution $\overline{\mathbf{W}}_0(t)$ of (2.37) lying in ${}^1\Omega_0$. Substituting $\overline{W}_{0i}(s)$ instead of $W_{0i}(s)$, $i = 1,\ldots,n$ into (2.37), we obtain the following differential equation:

$$
\begin{aligned}
g_i(t)W'_{0i}(t) = {} & a_i W_{0i}(t) + \left(a_i\varphi_i^{\mu}(t,C_i) + a_i W_{0i}(t)\right) \\
& \times \Bigg[f_i\Bigg(t, \varphi_1(t,C_1) + \varphi_1^{(1-\mu)}(t,C_1)(W_{01}(t) + T_{01}(t)),\ldots,\varphi_n(t,C_n) \\
& \quad + \varphi_n^{(1-\mu)}(t,C_n)\times(W_{0n}(t)+T_{0n}(t)), \\
& \quad \int_{0^+}^{t} K_i\Big(t,s,\varphi_1(t,C_1) + \varphi_1^{(1-\mu)}(t,C_1)(W_{01}(t)+T_{01}(t)),\ldots,\varphi_n(t,C_n) \\
& \qquad + \varphi_n^{(1-\mu)}(t,C_n)\times(W_{0n}(t)+T_{0n}(t)), \varphi_1(s,C_1) + \varphi_1^{(1-\mu)}(s,C_1) \\
& \qquad \times\left(\overline{W}_{01}(s) + T_{01}(s)\right),\ldots\varphi_n(s,C_n) \\
& \qquad + \varphi_n^{(1-\mu)}(s,C_n)\left(\overline{W}_{0n}(s) + T_{0n}(s)\right)\Big)ds\Bigg) \\
& - f_i\Bigg(t, \varphi_1(t,C_1) + \varphi_1^{(1-\mu)}(t,C_1)T_{01}(t),\ldots,\varphi_n(t,C_n) + \varphi_n^{(1-\mu)}(t,C_n), \\
& \quad \int_{0^+}^{t} K_i\Big(t,s,\varphi_1(t,C_1) + \varphi_1^{(1-\mu)}(t,C_1)T_{01}(t),\ldots,\varphi_n(t,C_n)T_{0n}(t) \\
& \qquad + \varphi_n^{(1-\mu)}(t,C_n)T_{0n}(t), \varphi_1(s,C_1) + \varphi_1^{(1-\mu)}(s,C_1) \\
& \qquad \times T_{01}(s),\ldots\varphi_n(s,C_n) + \varphi_n^{(1-\mu)}(s,C_n)T_{0n}(s)\Big)ds\Bigg)\Bigg].
\end{aligned}
\tag{2.40}
$$

Calculating the derivative $\dot{u}_{1i}(t,\mathbf{W}_0)$ along the trajectories of (2.40) on the set $\partial^1\Omega_0$, we get $\operatorname{sgn}\ \dot{u}_{1i}(t,\mathbf{W}_0) = -1$ for $t \in (0,t^{\Delta}]$, $i = 1,\ldots,n$.

By the same method as in the case of the existence of a solution of (1.1), we obtain that in ${}^1\Omega_0$ there is only the trivial solution of (2.40). The proof is complete. □

3. Asymptotic Expansions of Solutions

Diblík [3] investigated a singular initial problem for implicit ordinary differential equations and constructed asymptotic expansions of solutions in a right-hand neighbourhood of a singular point. Some results about asymptotic expansions of solutions for integrodifferential equations with separable kernels are given in [3, 10, 12].

The aim of this section is to show that results of paper [2] for ordinary differential equations are possible to extend on certain classes systems integrodifferential equations with a separable kernel in the following form:

$$g(t)y_i' = y_i + \int_{0^+}^{t} \left(\sum_{|\sigma_i|+|\omega_i|=2}^{N_i} u_{\sigma_i\omega_i}(t) v_{\sigma_i\omega_i}(s) \mathbf{y}^{\sigma_i}(t) \mathbf{y}^{\omega_i}(s) \right) ds, \tag{3.1}$$

where $N_i \in \mathbb{N}$, $\sigma_i = (l_{i1}, \ldots, l_{in})$, $\omega_i = (j_{i1}, \ldots, j_{in})$, $l_{ik}, j_{ik} \in \mathbb{N} \cup \{0\}$, $k = 1, \ldots, n$,

$$|\sigma_i| = \sum_{k=1}^{n} l_{ik}, \qquad |\omega_i| = \sum_{k=1}^{n} j_{ik}, \qquad \mathbf{y}^{\sigma_i}(t) = \prod_{k=1}^{n} y_k^{l_{ik}}(t), \qquad \mathbf{y}^{\omega_i}(s) = \prod_{k=1}^{n} y_k^{j_{ik}}(s),$$
$$u_{\sigma_i\omega_i}(t), v_{\sigma_i\omega_i}(t) \in C^0(J), \quad J = (0, t_0], \; i = 1, \ldots, n. \tag{3.2}$$

We will construct the solution of (3.1) in the form of one parametric asymptotic expansions as

$$y_i(t, C) = \sum_{h=1}^{\infty} f_{ih}(t) \phi^h(t, C), \quad i = 1, \ldots, n, \tag{3.3}$$

where $\phi(t, C)$ is the general solution of the differential equation $g(t)y' = y$ so that

$$\phi(t, C) = C \exp\left[\int_{t_0}^{t} \frac{d\tau}{g(\tau)} \right], \tag{3.4}$$

$f_{i1}(t) \equiv 1$, $f_{ih}(t)$, $h \geq 2$, $i = 1, \ldots, n$ are unknown functions, $C \neq 0$ is a constant.

Consider the following differential equation:

$$g(t)y' = qy + p(t). \tag{3.5}$$

Diblík [3] proved asymptotic estimates of the solution of (3.5) which we can be formulated as follows.

Theorem 3.1. *Assume that*

(I) *Let q be a constant, $q < 0$, $g(t) \in C^1(J)$, $g(t) > 0$, $\lim_{t \to t_0^+} g(t) = 0$, $g'(t) \sim \psi_1(t) g^{\lambda_1}(t)$ as $t \to t_0^+$, $\lambda_1 > 0$, $\lim_{t \to t_0^+} \psi_1(t) g^{\tau}(t) = 0$, τ is any positive number.*

(II) *$p(t) \in C(J)$, $p(t) = b_0(t) g^{\lambda}(t) + O(b_1(t) g^{\lambda+\epsilon}(t))$, $\epsilon > 0$, $\lim_{t \to t_0^+} b_m(t) g'(t) = 0$, $m = 0, 1$, $b_0(t) \in C(J)$, $b_0(t) \neq 0$, $b_0'(t) \sim \psi_2(t) g^{\lambda_2}(t)$ as $t \to t_0^+$, $\lambda_2 + 1 > 0$, $\lim_{t \to t_0^+} \psi_2(t) g^{\tau}(t) = 0$, $\lim_{t \to t_0^+} g^{\tau}(t) (b_0(t))^{-1} = 0$.*

Then (3.5) *has a unique solution on* $(0,\bar{t}]$, $0<\bar{t}\le t_0$, *satisfying asymptotic estimates*

$$y(x)=\frac{-1}{q}b_0(x)g^{\lambda}(x)+O(g^{\nu}(x)),\qquad y'(x)=O\left(g^{\nu-1}(x)\right),\tag{3.6}$$

where $\nu\in(\lambda,\lambda+\min\{\lambda_1,\lambda_2+1,\epsilon\})$.

Now we will show the results of Theorem 3.1. regarding only differential equation (3.5) we can apply to system of integrodifferential equations (3.1).

Substituting (3.3) into (3.1) and comparing the terms with the same powers of $\phi(t,C)$, we obtain the following system of recurrence equations:

$$g(t)f'_{ih}=(1-h)f_{ih}+\phi^{-h}(t,C)\int_{0^+}^{t}R_{ih}(t,s)ds,\tag{3.7}$$

$h\ge 2$, $i=1,\dots,n$ and

$$\begin{aligned}R_{ih}(t,s)=R_{ih}\big[&f_{11}(t),\dots,f_{ih-1}(t),\dots,f_{n1}(t),\dots,f_{nh-1}(t),\\&f_{11}(s),\dots,f_{1h-1}(s),\dots,f_{n1}(s),\dots,f_{nh-1}(s)\big].\end{aligned}\tag{3.8}$$

Denote

$$p_{ih}(t)=\phi^{-h}(t,C)\int_{0^+}^{t}R_{ih}(t,s)ds,\tag{3.9}$$

then it is obvious that the recurrence equations

$$g(t)f'_{ih}=(1-h)f_{ih}+p_{ih}(t)\tag{3.10}$$

$h\ge 2$, $i=1,\dots,n$ have the same form as (3.5) with the constant $q=1-h$. Hence we can apply Theorem 3.1, after the modification of assumption (II) of Theorem 3.1 for indices $h\ge 2$, $i=1,\dots,n$, to recurrence (3.10) which we will demonstrate with the following example.

Example 3.2. Consider the following system of integrodifferential equations:

$$\begin{aligned}t^2y_1'&=y_1+\int_{0^+}^{t}\frac{1}{t^3}y_1(s)y_2(s)ds,\\t^2y_2'&=y_2+\int_{0^+}^{t}\sqrt{t}y_1(t)y_2(s)ds.\end{aligned}\tag{3.11}$$

System (3.11) has the form of system (3.1) for

$$\begin{aligned}&\sigma_1\omega_1=(0,0,1,1),\qquad u_{\sigma_1\omega_1}(t)=\frac{1}{t^3},\qquad v_{\sigma_1\omega_1}(s)=1,\qquad N_1=2\\&\sigma_2\omega_2=(1,0,0,1),\qquad u_{\sigma_2\omega_2}(t)=\sqrt{t},\qquad v_{\sigma_2\omega_2}(s)=1,\qquad N_2=2.\end{aligned}\tag{3.12}$$

We will construct a solution of system (3.11) in the following form:

$$y_1 = \sum_{k=1}^{\infty} f_{1k}(t)\phi^k(t,C), \qquad y_2 = \sum_{k=1}^{\infty} f_{2k}(t)\phi^k(t,C), \tag{3.13}$$

where $\phi(t,C)$ is the general solution of the equation $t^2y' = y$. We will demonstrate the calculation of coefficients f_{ih} for $h = 3$. Substituting (3.13) in (3.11) and comparing the terms with the same powers of $\phi(t,C)$, we obtain the following system of recurrence equations:

$$\begin{aligned} \phi^1(t,C)&: 1 = 1, \\ &\quad 1 = 1. \end{aligned} \tag{3.14}$$

$$\begin{aligned} \phi^2(t,C)&: t^2 f_{12}' = -f_{12} + \phi^{-2}(t,C)\int_{0^+}^{t} \frac{1}{t^3}\phi^2(s,C)ds, \\ &\quad t^2 f_{22}' = -f_{22} + \phi^{-2}(t,C)\int_{0^+}^{t} \sqrt{t}\,\phi(t,C)\phi(s,C)ds. \end{aligned} \tag{3.15}$$

$$\begin{aligned} \phi^3(t,C)&: t^2 f_{13}' = -2f_{13} + \phi^{-3}(t,C)\int_{0^+}^{t} \frac{1}{t^3}[f_{12}(s) + f_{22}(s)]\phi^3(s,C)ds, \\ &\quad t^2 f_{23}' = -2f_{23} + \phi^{-3}(t,C)\int_{0^+}^{t} \sqrt{t}\Big[f_{12}(t)\phi^2(t,C)\phi(s,C) \\ &\qquad + f_{22}(s)\phi^2(s,C)\phi(t,C)\Big]ds. \end{aligned} \tag{3.16}$$

Put

$$u_1 = \phi^{-2}(t,C)\int_{0^+}^{t} \phi^2(s,C)ds, \qquad u_2 = \phi^{-1}(t,C)\int_{0^+}^{t} \phi(s,C)ds. \tag{3.17}$$

Differentiating both equations (3.17), we obtain the following differential equations:

$$t^2u_1' = -2u_1 + t^2, \tag{3.18}$$

$$t^2u_2' = -u_2 + t^2. \tag{3.19}$$

Equation (3.18) satisfies assumptions of Theorem 3.1. with following functions and coefficients:

$$\begin{aligned} &a = -2, \quad b_0(t) = 1, \quad g^{\lambda}(t) = \left(t^2\right)^1 \Rightarrow \lambda = 1, \quad b_1(t) = 0, \\ &g'(t) = \left(t^2\right)' = 2(g(t))^{1/2} \Rightarrow \lambda_1 = \frac{1}{2}, \quad b_0'(t) = 0 \cdot g^{\lambda_2}(t). \end{aligned} \tag{3.20}$$

Hence we can choose a constant $\lambda_2 + 1 > 1/2$ and similarly $\epsilon > 1/2$. By Theorem 3.1., we have

$$u_1 = \frac{1}{2}t^2 + O\left(t^{2\nu_1}\right), \qquad \nu_1 \in \left(1, \frac{3}{2}\right). \tag{3.21}$$

Second equation (3.19) is different from (3.18) only in the constant $a = -1$. Thus

$$u_2 = t^2 + O\left(t^{2\nu_2}\right), \qquad \nu_2 \in \left(1, \frac{3}{2}\right). \tag{3.22}$$

Substituting solutions (3.21) and (3.22) into (3.15) instead of integral terms, we obtain for unknown coefficients f_{12}, f_{22} the following differential equations:

$$t^2 f_{12}' = -f_{12} + \frac{1}{2t} + O\left(t^{2\nu_1 - 3}\right), \tag{3.23}$$

$$t^2 f_{22}' = -f_{22} + t^{5/2} + O\left(t^{2\nu_2 + 1/2}\right). \tag{3.24}$$

For (3.23), we can put

$$\begin{aligned} &a = -1, \qquad b_0(t) = \frac{1}{2}, \quad g^{\lambda}(t) = \left(t^2\right)^{-1/2}, \qquad \lambda = -\frac{1}{2}, \quad b_1(t) = 1, \\ &\epsilon = \nu_1 - 1, \qquad g'(t) = \left(t^2\right)' = 2\left(g(t)\right)^{1/2} \Rightarrow \lambda_1 = \frac{1}{2}, \qquad b_0'(t) = 0 \cdot g^{\lambda_2}(t). \end{aligned} \tag{3.25}$$

Then we can choose a constant $\lambda_2 + 1 > 1/2$. By Theorem 3.1., we get

$$f_{12}(t) = \frac{1}{2t} + O\left(t^{2\nu_{12}}\right), \qquad f_{12}'(t) = O\left(t^{2\nu_{12} - 2}\right), \qquad \nu_{12} \in \left(-\frac{1}{2}, 0\right). \tag{3.26}$$

Similarly for (3.24), we can put $a = -1$, $b_0(t) = 1$, $g^{\lambda}(t) = (t^2)^{5/4}$, $\lambda = 5/4, b_1(t) = 1$, $\epsilon = \nu_2 - 1$,

$$g'(t) = \left(t^2\right)' = 2\left(g(t)\right)^{1/2} \Longrightarrow \lambda_1 = \frac{1}{2}, \qquad b_0'(t) = 0 \cdot g^{\lambda_2}(t). \tag{3.27}$$

Then we can choose a constant $\lambda_2 + 1 > 1/2$. By Theorem 3.1., we have

$$f_{22}(t) = t^{5/2} + O\left(t^{2\nu_{22}}\right), \qquad f_{22}'(t) = O\left(t^{2\nu_{22} - 2}\right), \qquad \nu_{22} \in \left(\frac{5}{4}, \frac{7}{4}\right). \tag{3.28}$$

Substituting coefficients f_{12}, f_{22} into (3.16) and using the same method as in the calculation of coefficients f_{12}, f_{22}, we have

$$f_{13}(t) = \frac{1}{12t^2} + O\left(t^{2\nu_{13}}\right), \quad f_{13}'(t) = O\left(t^{2\nu_{13}-1}\right), \quad \nu_{13} \in \left(-1, -\frac{1}{2}\right),$$
$$f_{23}(t) = \frac{1}{4}t^{3/2} + O\left(t^{2\nu_{23}}\right), \quad f_{23}'(t) = O\left(t^{2\nu_{23}-1}\right), \quad \nu_{23} \in \left(\frac{3}{4}, \frac{5}{4}\right). \tag{3.29}$$

Thus the solution of system (3.11) has for $h = 3$ the following asymptotic expansions:

$$y_1 \approx \phi(t,C) + \left[\frac{1}{2t} + O\left(t^{2\nu_{12}}\right)\right]\phi^2(t,C) + \left[\frac{1}{12t^2} + O\left(t^{2\nu_{13}}\right)\right]\phi^3(t,C),$$
$$y_2 \approx \phi(t,C) + \left[t^{5/2} + O\left(t^{2\nu_{22}}\right)\right]\phi^2(t,C) + \left[\frac{1}{4}t^{3/2} + O\left(t^{2\nu_{23}}\right)\right]\phi^3(t,C). \tag{3.30}$$

Acknowledgments

The first author is supported by Grant FEKT-S-11-2-921 of the Faculty of Electrical Engineering and Communication, Brno University of Technology and Grant P201/11/0768 of the Czech Grant Agency (Prague).

References

[1] R. P. Agarwal, D. O'Regan, and O. E. Zernov, "A singular initial value problem for some functional differential equations," *Journal of Applied Mathematics and Stochastic Analysis*, vol. 2004, no. 3, pp. 261–270, 2004.

[2] V. A. Čečik, "Investigation of systems of ordinary differential equations with a singularity," *Trudy Moskovskogo Matematičeskogo Obščestva*, vol. 8, pp. 155–198, 1959 (Russian).

[3] I. Diblík, "Asymptotic behavior of solutions of a differential equation partially solved with respect to the derivative," *Siberian Mathematical Journal*, vol. 23, no. 5, pp. 654–662, 1982 (Russian).

[4] J. Diblík, "Existence of solutions of a real system of ordinary differential equations entering into a singular point," *Ukrainian Mathematical Journal*, vol. 38, no. 6, pp. 588–592, 1986 (Russian).

[5] J. Baštinec and J. Diblík, "On existence of solutions of a singular Cauchy-Nicoletti problem for a system of integro-differential equations," *Demonstratio Mathematica*, vol. 30, no. 4, pp. 747–760, 1997.

[6] J. Diblík, "On the existence of $\sum_{k=1}^{n}(a_{k1}t+a_{k2}x)(x')^k = b_1t+b_2x+f(t,x,x'), x(0) = 0$-curves of a singular system of differential equations," *Mathematische Nachrichten*, vol. 122, pp. 247–258, 1985 (Russian).

[7] J. Diblík and C. Nowak, "A nonuniqueness criterion for a singular system of two ordinary differential equations," *Nonlinear Analysis. Theory, Methods & Applications A*, vol. 64, no. 4, pp. 637–656, 2006.

[8] J. Diblík and M. Růžičková, "Existence of positive solutions of a singular initial problem for a nonlinear system of differential equations," *The Rocky Mountain Journal of Mathematics*, vol. 34, no. 3, pp. 923–944, 2004.

[9] J. Diblík and M. R. Růžičková, "Inequalities for solutions of singular initial problems for Caratheodory systems via Ważewski's principle," *Nonlinear Analysis: Theory, Methods and Applications*, vol. 69, no. 12, pp. 657–656, 2008.

[10] Z. Šmarda, "On the uniqueness of solutions of the singular problem for certain class of integro-differential equations," *Demonstratio Mathematica*, vol. 25, no. 4, pp. 835–841, 1992.

[11] Z. Šmarda, "On a singular initial value problem for a system of integro-differential equations depending on a parameter," *Fasciculi Mathematici*, no. 25, pp. 123–126, 1995.

[12] Z. Šmarda, "On an initial value problem for singular integro-differential equations," *Demonstratio Mathematica*, vol. 35, no. 4, pp. 803–811, 2002.

[13] Z. Šmarda, "Implicit singular integrodifferential equations of Fredholm type," *Tatra Mountains Mathematical Publications*, vol. 38, pp. 255–263, 2007.

[14] A. E. Zernov and Yu. V. Kuzina, "Qualitative investigation of the singular Cauchy problem $\sum_{k=1}^{n}(a_{k1}t + a_{k2}x)(x')^k = b_1 t + b_2 x + f(t, x, x'), x(0) = 0$," *Ukrainian Mathematical Journal*, vol. 55, no. 10, pp. 1419–1424, 2003 (Russian).

[15] A. E. Zernov and Yu. V. Kuzina, "Geometric analysis of a singular Cauchy problem," *Nonlinear Oscillations*, vol. 7, no. 1, pp. 67–80, 2004 (Russian).

[16] A. E. Zernov and O. R. Chaĭchuk, "Asymptotic behavior of solutions of a singular Cauchy problem for a functional-differential equation," *Journal of Mathematical Sciences*, vol. 160, no. 1, pp. 123–135, 2009.

[17] R. Srzednicki, "Ważewski method and Conley index," in *Handbook of Differential Equations: Ordinary Differential Equations*, A. Canada, P. Drabek, and A. Fonda, Eds., vol. 1, pp. 591–684, Elsevier, Amsterdam, The Netherlands, 2004.

[18] P. Hartman, *Ordinary Differential Equations*, John Wiley & Sons, New York, NY, USA, 1964.

[19] E. Zeidler, *Applied Functional Analysis: Applications to Mathematical Physics*, vol. 108 of *Applied Mathematical Sciences*, Springer, New York, NY, USA, 1999.

6

On Some Properties of New Paranormed Sequence Space of Nonabsolute Type

Vatan Karakaya[1] and Necip Şimşek[2]

[1] *Department of Mathematical Engineering, Yildiz Technical University, Davutpasa Campus, Esenler 34220, Istanbul, Turkey*
[2] *Department of Mathematic, Istanbul Commerce University, Üsküdar, Istanbul, Turkey*

Correspondence should be addressed to Vatan Karakaya, vkkaya@yildiz.edu.tr

Academic Editor: Qiji J. Zhu

We introduce some new generalized sequence space related to the space $\ell(p)$. Furthermore we investigate some topological properties as the completeness, the isomorphism, and also we give some inclusion relations between this sequence space and some of the other sequence spaces. In addition, we compute α-, β-, and γ-duals of this space and characterize certain matrix transformations on this sequence space.

1. Introduction

In studying the sequence spaces, especially, to obtain new sequence spaces, in general, the matrix domain μ_A of an infinite matrix A defined by $\mu_A = \{x = (x_k) \in w : Ax \in \mu\}$ is used. In most cases, the new sequence space μ_A generated by a sequence space μ is the expansion or the contraction of the original space μ. In some cases, these spaces could be overlapped. Indeed, one can easily see that the inclusion $\mu_S \subset \mu$ strictly holds for $\mu \in \{\ell_\infty, c, c_0\}$. Similarly one can deduce that the inclusion $\mu \subset \mu_\Delta$ also strictly holds for $\mu \in \{\ell_\infty, c, c_0\}$, where S and Δ are matrix operators.

Recently, in [1], Mursaleen and Noman constructed new sequence spaces by using matrix domain over a normed space. They also studied some topological properties and inclusion relations of these spaces.

It is well known that paranormed spaces have more general properties than the normed spaces. In this work, we generalize the normed sequence spaces defined by Mursaleen and Noman [1] to the paranormed spaces. Furthermore we introduce new sequence space over the paranormed space. Next we investigate behaviors of this sequence space according to topological properties and inclusion relations. Finally we give certain matrix transformation on this sequence space and its duals.

In the literature, by using the matrix domain over the paranormed spaces, many authors have defined new sequence spaces. Some of them are as follows. For example, Choudhary and Mishra [2] have defined the sequence space $\overline{\ell(p)}$ where the S-transform is in $\ell(p)$, Başar and Altay [3, 4] defined the spaces $\lambda(u,v;p) = \{\lambda(p)\}_G$ for $\lambda \in \{\ell_\infty, c, c_0\}$ and $\ell(u,v;p) = \{\ell(p)\}_G$, respectively, and Altay and Başar [5] have defined the spaces $r^t_\infty(p)$, $r^t_c(p)$, $r^t_0(p)$. In [6], Karakaya and Polat defined and examined the spaces $e^r_0(\Delta;p)$, $e^r(\Delta;p)$, $e^r_\infty(\Delta;p)$, and Karakaya et al. [7] have recently introduced and studied the spaces $\ell_\infty(\lambda,p)$, $c(\lambda,p)$, $c_0(\lambda,p)$, where R^t and E^r denote the Riesz and the Euler means, respectively, Δ denotes the band matrix of the difference operators, and Λ, G are defined in [1, 8], respectively. Also, the information on matrix domain of sequence spaces can be found in [9–13].

By w, we denote the space of all real valued sequences. Any vector subspace of w is called a sequence space. By the spaces ℓ_1, cs, and bs, we denote the spaces of all absolutely convergent series, convergent series, and bounded series, respectively.

A linear topological space X over the real field $\mathbb{R}$ is said to be a paranormed space if there is a subadditivity function $h : X \to \mathbb{R}$ such that $h(\theta) = 0$, $h(x) = h(-x)$, and scalar multiplication is continuous, that is, $|\alpha_n - \alpha| \to 0$ and $h(x_n - x) \to 0$ imply $h(\alpha_n x_n - \alpha x) \to 0$ for all α in $\mathbb{R}$ and x in X, where θ is the zero in the linear space X.

Let μ, ν be any two sequence spaces, and let $A = (a_{nk})$ be any infinite matrix of real number a_{nk}, where $n, k \in \mathbb{N}$ with $\mathbb{N} = \{0, 1, 2, \ldots\}$. Then we say that A defines a matrix mapping from μ into ν by writing$A : \mu \to \nu$, if for every sequence $x = (x_k) \in \mu$, the sequence $Ax = (A_n(x))$, the A-transform of x, is in ν, where

$$A_n(x) = \sum_k a_{nk} x_k \quad (n \in \mathbb{N}). \tag{1.1}$$

By (μ, ν), we denote the class of all matrices A such that $A : \mu \to \nu$. Thus, $A \in (\mu, \nu)$ if and only if the series on the right hand side of (1.1) converges for each $n \in \mathbb{N}$ and every $x \in \mu$, and we have $Ax \in \nu$ for all $x \in \mu$. A sequence x is said to be A-summable to a if Ax converges to a which is called as the A-limit of x.

Assume here and after that (p_k), (q_k) are bounded sequences of strictly positive real numbers with sup $p_k = H$ and $M = \max(1, H)$, and also let $p'_k = p_k/(p_k - 1)$ for $1 < p_k < \infty$ and for all $k \in \mathbb{N}$. The linear space $\ell(p)$ was defined by Maddox [14] as follows:

$$\ell(p) = \left\{ x = (x_n) \in w : \sum_{n=0}^{\infty} |x_n|^{p_n} < \infty \right\} \tag{1.2}$$

which are the complete space paranormed by

$$h(x) = \left(\sum_{n=0}^{\infty} |x_n|^{p_n} \right)^{1/M}. \tag{1.3}$$

Throughout this work, by $\mathcal{F}$ and N_k, respectively, we will denote the collection of all subsets of $\mathbb{N}$ and the set of all $n \in \mathbb{N}$ such that $n \geq k$ and $e = (1, 1, 1, \ldots)$.

2. The Sequence Space $\ell(\lambda, p)$

In this section, we define the sequence space $\ell(\lambda, p)$ and prove that this sequence space according to its paranorm is complete paranormed linear space. In [1], Mursaleen and Noman defined the matrix $\Lambda = (\lambda_{nk})_{n,k=0}^{\infty}$ by

$$\lambda_{nk} = \begin{cases} \dfrac{\lambda_k - \lambda_{k-1}}{\lambda_n} & (0 \leq k \leq n) \\ 0 & (k > n), \end{cases} \tag{2.1}$$

where $\lambda = (\lambda_k)_{k=0}^{\infty}$ is a strictly increasing sequence of positive reals tending to ∞, that is, $0 < \lambda_0 < \lambda_1 < \cdots$ and $\lambda_k \to \infty$ as $k \to \infty$. Now, by using (2.1) we define new sequence space as follows:

$$\ell(\lambda, p) = \left\{ x = (x_k) \in w : \sum_{n=0}^{\infty} \left| \frac{1}{\lambda_n} \sum_{k=0}^{n} (\lambda_k - \lambda_{k-1}) x_k \right|^{p_n} < \infty \right\}. \tag{2.2}$$

For any $x = (x_n) \in w$, we define the sequence $y = (y_n)$, which will frequently be used, as the Λ-transform of x, that is, $y = \Lambda(x)$, and hence

$$y_n = \sum_{k=0}^{n} \left(\frac{\lambda_k - \lambda_{k-1}}{\lambda_n} \right) x_k \quad (n \in N). \tag{2.3}$$

We now may begin with the following theorem.

Theorem 2.1. *The sequence space $\ell(\lambda, p)$ is the complete linear metric space with respect to paranorm defined by*

$$h(x) = \left(\sum_{n=0}^{\infty} \left| \frac{1}{\lambda_n} \sum_{k=0}^{n} (\lambda_k - \lambda_{k-1}) x_k \right|^{p_n} \right)^{1/M}. \tag{2.4}$$

Proof. The linearity of $\ell(\lambda, p)$ with respect to the coordinatewise addition and scalar multiplication follows from the following inequalities which are satisfied for $x, t \in \ell(\lambda, p)$ (see, [15]):

$$\begin{aligned} \left(\sum_{n=0}^{\infty} \left| \frac{1}{\lambda_n} \sum_{k=0}^{n} (\lambda_k - \lambda_{k-1})(x_k + t_k) \right|^{p_n} \right)^{1/M} &\leq \left(\sum_{n=0}^{\infty} \left| \frac{1}{\lambda_n} \sum_{k=0}^{n} (\lambda_k - \lambda_{k-1}) x_k \right|^{p_n} \right)^{1/M} \\ &\quad + \left(\sum_{n=0}^{\infty} \left| \frac{1}{\lambda_n} \sum_{k=0}^{n} (\lambda_k - \lambda_{k-1}) t_k \right|^{p_n} \right)^{1/M} \end{aligned} \tag{2.5}$$

and for any $\alpha \in \mathbb{R}$ (see, [16])

$$|\alpha|^{p_k} \leq \max\left\{1, |\alpha|^M\right\}. \tag{2.6}$$

It is clear that $h(\theta) = 0$, $h(x) = h(-x)$ for all $x \in \ell(\lambda, p)$. Again inequalities (2.5) and (2.6) yield the subadditivity of h and hence $h(\alpha x) \leq \max\{1, |\alpha|^M\} h(x)$. Let $\{x^m\}$ be any sequence of points $x^m \in \ell(\lambda, p)$ such that $h(x^m - x) \to 0$ and (α_m) also any sequence of scalars such that $\alpha_m \to \alpha$. Then, since the inequality

$$h(x^m) \leq h(x) + h(x^m - x) \tag{2.7}$$

holds by subadditivity of h, we can write that $\{h(x^m)\}$ is bounded and we thus have

$$\begin{aligned} h(\alpha_m x^m - \alpha x) &= \left(\sum_{n=0}^{\infty} \left| \frac{1}{\lambda_n} \sum_{k=0}^{n} (\lambda_k - \lambda_{k-1})(\alpha_m x_k^m - \alpha x_k) \right|^{p_n} \right)^{1/M} \\ &\leq |\alpha_m \longrightarrow \alpha|^{1/M} h(x^m) + |\alpha|^{1/M} h(x^m - x) \end{aligned} \tag{2.8}$$

which tends to zero as $n \to \infty$. Therefore, the scalar multiplication is continuous. Hence h is a paranorm on the space $\ell(\lambda, p)$. It remains to prove the completeness of the space $\ell(\lambda, p)$. Let $\{x^j\}$ be any Cauchy sequence in the space $\ell(\lambda, p)$, where $x^j = \{x_0^{(j)}, x_1^{(j)}, x_2^{(j)}, \ldots\}$. Then, for a given $\varepsilon > 0$, there exists a positive integer $m_0(\varepsilon)$ such that $h(x^j - x^i) < \varepsilon/2$ for all $i, j > m_0(\varepsilon)$. Using definition of h, we obtain for each fixed $n \in \mathbb{N}$ that

$$\left| \Lambda_n\left(x^j\right) - \Lambda_n\left(x^i\right) \right| \leq \left(\sum_{n=0}^{\infty} \left| \Lambda_n\left(x^j\right) - \Lambda_n\left(x^i\right) \right|^{p_n} \right)^{1/M} < \frac{\varepsilon}{2} \tag{2.9}$$

for every $i,j > m_0(\varepsilon)$ which leads us to the fact that $\{\Lambda_n(x^0), \Lambda_n(x^1), \Lambda_n(x^2), \ldots\}$ is a Cauchy sequence of real numbers for every fixed $n \in \mathbb{N}$. Since $\mathbb{R}$ is complete, it converges, say $\Lambda_n(x^i) - \Lambda_n(x)$ as $i \to \infty$. Using these infinitely many limits, we may write the sequence $\{\Lambda_0(x), \Lambda_1(x), \Lambda_2(x), \ldots\}$. From (2.9) as $i \to \infty$, we have

$$\left| \Lambda_n\left(x^j\right) - \Lambda_n(x) \right| < \frac{\varepsilon}{2}, \quad (j \geq m_0(\varepsilon)) \tag{2.10}$$

for every fixed $n \in \mathbb{N}$. By using (2.9) and boundedness of Cauchy sequence, we have

$$\left(\sum_{n=0}^{\infty} |\Lambda_n(x)|^{p_n} \right)^{1/M} \leq \left(\sum_{n=0}^{\infty} \left| \Lambda_n\left(x^j\right) - \Lambda_n\left(x^i\right) \right|^{p_n} \right)^{1/M} + \left(\sum_{n=0}^{\infty} \left| \Lambda_n\left(x^j\right) \right|^{p_n} \right)^{1/M} < \infty. \tag{2.11}$$

Hence, we get $x \in \ell(\lambda, p)$. So, the space $\ell(\lambda, p)$ is complete. □

Theorem 2.2. *The sequence space $\ell(\lambda, p)$ of nonabsolute type is linearly isomorphic to the space $\ell(p)$, where $0 < p_k \leq H < \infty$.*

Proof. To prove the theorem, we would show the existence of linear bijection between the spaces $\ell(\lambda, p)$ and $\ell(p)$. With the notation of (2.3), we define transformation T from $\ell(\lambda, p)$ to $\ell(p)$ by $x \to y = Tx$. The linearity of T is trivial. Furthermore, it is obvious that $x = \theta$ whenever $Tx = \theta$ and hence T is injective.

Let $y \in \ell(p)$ and define the sequence $x = \{x_n\}$

$$x_n(\lambda) = \sum_{k=n-1}^{n} \left((-1)^{n-k} \frac{\lambda_k}{\lambda_n - \lambda_{n-1}} \right) y_k \quad (n, k \in \mathbb{N}). \tag{2.12}$$

Then, we have

$$h_{\ell(\lambda,p)}(x) = \left(\sum_{n=0}^{\infty} \left| \frac{1}{\lambda_n} \sum_{k=0}^{n} (\lambda_k - \lambda_{k-1}) x_k \right|^{p_n} \right)^{1/M} = \left(\sum_{n=0}^{\infty} |y_n|^{p_n} \right)^{1/M} = h_{\ell(p)}(y). \tag{2.13}$$

Thus, we have that $x \in \ell(\lambda, p)$ and consequently T is surjective. Hence, T is a linear bijection and this tells us that the spaces $\ell(\lambda, p)$ and $\ell(p)$ are linearly isomorphic. This completes the proof. □

3. Some Inclusion Relations

In this section, we give some inclusion relations concerning the space $\ell(\lambda, p)$. Before giving the theorems about the section, we give a lemma given in [1].

Lemma 3.1. *For any sequence $x = (x_k) \in w$, the equalites*

$$S_n(x) = x_n - \Lambda_n(x), \tag{3.1}$$

$$S_n(x) = \frac{\lambda_{n-1}}{\lambda_n - \lambda_{n-1}} [\Lambda_n(x) - \Lambda_{n-1}(x)] \tag{3.2}$$

hold, where the sequence $S(x) = \{S_n(x)\}$ is defined by

$$S_0(x) = 0, \qquad S_n(x) = \frac{1}{\lambda_n} \sum_{k=1}^{n} \lambda_{k-1}(x_k - x_{k-1}) \quad (n \geq 1). \tag{3.3}$$

Theorem 3.2. *The inclusion $\ell(\lambda, p) \subset c_0(\lambda, p)$ holds.*

Proof. Let $x \in \ell(\lambda, p)$. It can be written $\Lambda x \in \ell(p)$. By the definition of the space $\ell(p)$, $\Lambda_n x \to \infty$ as $n \to \infty$, we obtain $\Lambda x \in c_0$. Hence we get $x \in c_0(\lambda, p)$. □

Theorem 3.3. *The inclusion $\ell(\lambda, p) \subset \ell(p)$ if and only if $S(x) \in \ell(p)$ for every sequence $x \in \ell(\lambda, p)$, where $1 \leq p_k \leq H$.*

Proof. We suppose that $\ell(\lambda, p) \subset \ell(p)$ holds and take any $x \in \ell(\lambda, p)$. Then $x \in \ell(p)$ by hypothesis. Thus we obtain from (3.1) that

$$[h(S(x))]_{\ell(p)} \leq [h(x)]_{\ell(p)} + [h(\Lambda x)]_{\ell(p)} = [h(x)]_{\ell(p)} + [h(x)]_{\ell(\lambda,p)} \tag{3.4}$$

which yields that $S(x) \in \ell(p)$.

Conversely, let $x \in \ell(\lambda, p)$ be given. Then we have by the hypothesis that $S(x) \in \ell(p)$. Again by using (3.1)

$$[h(x)]_{\ell(p)} \leq [h(S(x))]_{\ell(p)} + [h(\Lambda x)]_{\ell(p)} = [h(S(x))]_{\ell(p)} + [h(x)]_{\ell(\lambda,p)} \tag{3.5}$$

which shows that $x \in \ell(p)$. Hence the inclusion $\ell(\lambda, p) \subset \ell(p)$ holds. This completes the proof. □

Theorem 3.4.

(i) If $p_n > 1$ for all $n \in N$, then the inclusion $\ell_p^\lambda \subset \ell(\lambda, p)$ holds.

(ii) If $p_n < 1$ for all $n \in N$, then the inclusion $\ell(\lambda, p) \subset \ell_p^\lambda$ holds.

Proof. (i) If $p = (p_n)$ for all $n \in N$, then we write ℓ_p^λ in place of$\ell(\lambda, p)$. Let $x \in \ell_p^\lambda$. It is clear that $\Lambda(x) \in \ell_p$. One can find $m \in N$ such that $|\Lambda_n(x)| < 1$ for all $n \geq m$. Under condition (i), we have $|\Lambda_n(x)|^{p_n} < |\Lambda_n(x)|$ for all $n \geq m$. Hence we get $x \in \ell(\lambda, p)$.

(ii) We suppose that $x \in \ell(\lambda, p)$. Then $\Lambda(x) \in \ell(p)$ and there exists $m \in N$ such that $|\Lambda_n(x)|^{p_n} < 1$ for all $n \geq m$. To obtain the result, we consider the following inequality:

$$|\Lambda_n(x)| = \left(|\Lambda_n(x)|^{p_n}\right)^{1/p_n} < |\Lambda_n(x)|^{p_n} \tag{3.6}$$

for all $n \geq m$. So, we get $x \in \ell_p^\lambda$. □

4. Some Matrix Transformations and Duals of the Space $\ell(\lambda, p)$

In this section, we give the theorems determining the α-, β-, and γ-duals of the space $\ell(\lambda, p)$. In proving the theorem, we apply the technique used in [3]. Also we give some matrix transformations from the space $\ell(\lambda, p)$ into paranormed spaces $\ell(q)$ by using the matrix given in [1].

For the sequence space μ and ν, the set $S(\mu, \nu)$ defined by

$$S(\mu, \nu) = \{a = (a_k) \in w : ax \in \nu \ \forall x \in \mu\} \tag{4.1}$$

is called the multiplier space of μ and ν. The α-, β-, and γ-duals of a sequence space μ, which are, respectively, denoted by μ^α, μ^β, and μ^γ, are defined by

$$\mu^\alpha = S(\mu, \ell_1), \qquad \mu^\beta = S(\mu, cs), \qquad \mu^\gamma = S(\mu, bs). \tag{4.2}$$

We may begin with the following theorem which computes the α-dual of the space $\ell(\lambda, p)$.

Theorem 4.1. *Let $K_1 = \{k \in \mathbb{N} : p_k \leq 1\}$ and $K_2 = \{k \in \mathbb{N} : p_k > 1\}$. Define the matrix $D^a = (d_{nk}^a)$ by*

$$d_{nk}^a = \begin{cases} (-1)^{n-k} \dfrac{\lambda_k}{\lambda_n - \lambda_{n-1}} a_n, & (n-1 \leq k \leq n) \\ 0, & (0 \leq k \leq n-1) \text{ or } (k > n). \end{cases} \tag{4.3}$$

Then

$$\begin{aligned} \ell_{K_1}^{\alpha}(\lambda,p) &= \{a=(a_n)\in w : D^a \in (\ell(p);\ell_\infty)\} \\ \ell_{K_2}^{\alpha}(\lambda,p) &= \{a=(a_n)\in w : D^a \in (\ell(p);\ell_1)\}. \end{aligned} \tag{4.4}$$

Proof. We consider the following equality:

$$a_n x_n = \sum_{k=n-1}^{n} d_{nk}^a y_k = (D^a y)_n \quad (n \in \mathbb{N}), \tag{4.5}$$

where $D^a = (d_{nk}^a)$ is defined by (4.3).

From (4.5), it can be obtained that $ax = (a_n x_n) \in \ell_1$ or $ax = (a_n x_n) \in \ell_\infty$ whenever $x \in \ell(\lambda,p)$ if and only if $D^a y \in \ell_1$ or $D^a y \in \ell_\infty$ whenever $y \in \ell(p)$.This means $a \in \ell_{K_1}^{\alpha}(\lambda,p)$ or $a \in \ell_{K_2}^{\alpha}(\lambda,p)$ if and only if $D^a \in (\ell(p);\ell_1)$ or $D^a \in (\ell(p);\ell_\infty)$. Hence this completes the proof. □

The result of the Theorem above corresponds the Theorem 5.1(0, 8, 12) given in [17].

As a direct consequence of Theorem 4.1, we have the following.

Corollary 4.2. *Let* $K^* = \{k \in \mathbb{N} : n-1 \le k \le n\} \cap K$ *for* $K \in \mathcal{F}$. *Then*

(i) $\ell_{K_1}^{\alpha}(\lambda,p) = \{a=(a_n)\in w : \sup_N \sup_{k\in\mathbb{N}} |\sum_{n\in K^*} d_{nk}^a|^{p_k} < \infty\}$,

(ii) $\ell_{K_2}^{\alpha}(\lambda,p) = \bigcup_{M>1}\{a=(a_n)\in w : \sup_{K\in\mathcal{F}} \sum_k |\sum_{n\in K^*} d_{nk}^a M^{-1}|^{p_k'} < \infty\}$

In the following theorem, we characterize the β- and γ-duals of the space $\ell(\lambda,p)$.

Theorem 4.3. *Let* $K_1 = \{k \in \mathbb{N} : p_k \le 1\}$, $K_2 = \{k \in \mathbb{N} : p_k > 1\}$, *and let* $\Delta x_k = x_k - x_{k+1}$. *Define the sequence* $s^1 = (s_k^1)$, $s^2 = (s_k^2)$ *and the matrix* $B^a = (b_{nk}^a)$ *by*

$$\begin{gathered} s_k^1 = \Delta\left(\frac{a_k}{\lambda_k - \lambda_{k-1}}\right)\lambda_k, \qquad s_k^2 = \frac{a_k \lambda_k}{\lambda_k - \lambda_{k-1}} \\ b_{nk}^a = \begin{cases} s_k^1, & (0 \le k \le n-1) \\ s_k^2, & (k = n) \\ 0, & (k > n) \end{cases} \end{gathered} \tag{4.6}$$

for all $n,k \in \mathbb{N}$. *Then*

$$\ell_{K_1}^{\beta}(\lambda,p) = \ell_{K_1}^{\gamma}(\lambda,p) = \{a=(a_n)\in w : B^a \in (\ell(p);\ell_\infty)\}, \tag{4.7}$$

$$\ell_{K_2}^{\beta}(\lambda,p) = \ell_{K_2}^{\gamma}(\lambda,p) = \{a=(a_n)\in w : B^a \in (\ell(p);c)\}. \tag{4.8}$$

Proof. Consider the equality

$$\sum_{k=0}^{n} a_k x_k = \sum_{k=0}^{n-1} s_k^1 y_k + s_n^2 yn = \left(B^a y\right)_n. \tag{4.9}$$

From (4.9), it can be obtained that $ax = (a_n x_n) \in cs$ or bs whenever $x = (x_n) \in \ell(\lambda, p)$ if and only if $B^a y \in c$ or ℓ_∞ whenever $y = (y_k) \in \ell(p)$. This means that $a = (a_n) \in \{\ell_{K_1}^{\beta}(\lambda, p)$ or $\ell_{K_2}^{\beta}(\lambda, p)\}$ or $a = (a_n) \in \{\ell_{K_1}^{\gamma}(\lambda, p)$ or $\ell_{K_2}^{\gamma}(\lambda, p)\}$ if and only if $B^a \in (\ell(p); c)$ or $B^a \in (\ell(p); \ell_\infty)$. Hence this completes the proof. □

We can write the following corollary from Theorem 4.3.

Corollary 4.4. *Let* $p'_k = p_k/(p_k - 1)$ *for* $1 < p_k < \infty$ *and for all* $k \in \mathbb{N}$. *Then*

(i) $\ell_{K_1}^{\beta}(\lambda, p) = \ell_{K_1}^{\gamma}(\lambda, p) = \{a = (a_n) \in w : s^1, s^2 \in \ell_\infty(p)\}$,

(ii) $\ell_{K_2}^{\beta}(\lambda, p) = \ell_{K_2}^{\gamma}(\lambda, p) = \bigcup_{M>1}\{a = (a_n) \in w : s^1 M^{-1}, s^2 M^{-1} \in \ell(p') \cap \ell_\infty(p')\}$.

After this step, we can give our theorems on the characterization of some matrix classes concerning the sequence space $\ell(\lambda, p)$.

Let $x, y \in w$ be connected by the relation $y = \Lambda(x)$. For an infinite matrix $A = (a_{nk})$, we have by using (4.9) of Theorem 4.3 that

$$\sum_{k=0}^{m} a_{nk} x_k = \sum_{k=0}^{m-1} \tilde{a}_{nk} y_k + \frac{\lambda_m}{\lambda_m - \lambda_{m-1}} a_{nm} y_m \quad (m, n \in \mathbb{N}), \tag{4.10}$$

where

$$\tilde{a}_{nk} = \left(\frac{a_{nk}}{\lambda_k - \lambda_{k-1}} - \frac{a_{n,k+1}}{\lambda_{k+1} - \lambda_k}\right)\lambda_k \quad (n, k \in \mathbb{N}). \tag{4.11}$$

The necessary and sufficient conditions characterizing the matrix mapping of the sequence space $\ell(p)$ of Maddox have been determined by Grosse-Erdmann [17]. Let L and M be the natural numbers and define the sets by $K_1 = \{k \in \mathbb{N} : p_k \leq 1\}$ and $K_2 = \{k \in \mathbb{N} : p_k > 1\}$ and also let us put $p'_k = p_k/(p_k - 1)$ for $1 < p_k < \infty$ and for all $k \in \mathbb{N}$. Before giving the theorems, let us suppose that (q_n) is a nondecreasing bounded sequence of positive real numbers and consider the following conditions:

$$\sup_N \sup_{k\in K_1} \left| \sum_{n\in N} \widetilde{a}_{nk} \right|^{q_n} < \infty, \tag{4.12}$$

$$\exists M \sup_N \sum_{k\in K_2} \left|_{n\in N} \widetilde{a}_{nk} M^{-1} \right|^{p'_k} < \infty, \tag{4.13}$$

$$\exists M \sup_k \sum_n \left| \widetilde{a}_k M^{-1/p_k} \right|^{q_n} < \infty, \tag{4.14}$$

$$\lim_n |\widetilde{a}_{nk}|^{q_n} = 0 \quad (\forall k \in \mathbb{N}), \tag{4.15}$$

$$\forall L, \sup_n \sup_{k\in K_1} \left| \widetilde{a}_{nk} L^{1/q_n} \right|^{p_k} < \infty, \tag{4.16}$$

$$\forall L, \exists M \sup_n \sum_{k\in K_2} \left| \widetilde{a}_k L^{1/q_n} M^{-1} \right|^{p'_k} < \infty, \tag{4.17}$$

$$\sup_n \sup_{k\in K_1} |\widetilde{a}_{nk}|^{p_k} < \infty, \tag{4.18}$$

$$\exists M \sup_n \sum_{k\in K_2} \left| \widetilde{a}_k M^{-1} \right|^{p'_k} < \infty, \tag{4.19}$$

$$\forall L, \sup_n \sup_{k\in K_1} \left(|\widetilde{a}_{nk} - \widetilde{a}_k| L^{1/q_n} \right)^{p_k} < \infty, \tag{4.20}$$

$$\lim_n |\widetilde{a}_{nk} - \widetilde{a}_k|^{q_n} = 0, \quad \forall k. \tag{4.21}$$

$$\forall L, \exists M \sup_n \sum_{k\in K_2} \left(|\widetilde{a}_{nk} - \widetilde{a}_k| L^{1/q_n} M^{-1} \right)^{p'_k}, \tag{4.22}$$

$$\exists L, \sup_n \sup_{k\in K_1} \left| \widetilde{a}_{nk} L^{-1/q_n} \right|^{p_k} < \infty, \tag{4.23}$$

$$\exists L, \sup_n \sum_{k\in K_2} \left| \widetilde{a}_{nk} L^{-1/q_n} \right|^{p'_k} < \infty, \tag{4.24}$$

$$\left(\frac{\lambda_k}{\lambda_k - \lambda_{k-1}} a_{nk} \right)_{k=0}^{\infty} \in c_0(q) \quad (\forall n \in \mathbb{N}) \tag{4.25}$$

$$\left(\frac{\lambda_k}{\lambda_k - \lambda_{k-1}} a_{nk} \right)_{k=0}^{\infty} \in c(q) \quad (\forall n \in \mathbb{N}) \tag{4.26}$$

$$\left(\frac{\lambda_k}{\lambda_k - \lambda_{k-1}} a_{nk} \right)_{k=0}^{\infty} \in \ell_\infty(q) \quad (\forall n \in \mathbb{N}). \tag{4.27}$$

By using (4.7), (4.10), and Corollary 4.4, we have the following results.

Theorem 4.5. *One has the following:*

(i) $A \in (\ell(\lambda, p) : \ell(q))$ *if and only if* (4.12), (4.13), (4.14), *and* (4.25) *hold,*

(ii) $A \in (\ell(\lambda, p) : c_0(q))$ *if and only if* (4.15), (4.16), (4.17), *and* (4.25) *hold,*

(iii) $A \in (\ell(\lambda, p) : c(q))$ *if and only if* (4.18), (4.19), (4.20), (4.21), (4.22), *and* (4.26) *hold,*

(iv) $A \in (\ell(\lambda, p) : \ell_\infty(q))$ *if and only if* (4.23), (4.24), *and* (4.27) *hold.*

References

[1] M. Mursaleen and A. K. Noman, "On the spaces of λ-convergent and bounded sequences," *Thai Journal of Mathematics*, vol. 8, no. 2, pp. 311–329, 2010.

[2] B. Choudhary and S. K. Mishra, "On Köthe-Toeplitz duals of certain sequence spaces and their matrix transformations," *Indian Journal of Pure and Applied Mathematics*, vol. 24, no. 5, pp. 291–301, 1993.

[3] F. Başar and B. Altay, "Some paranormed sequence spaces of non-absolute type derived by weighted mean," *Journal of Mathematical Analysis and Applications*, vol. 319, no. 2, pp. 494–508, 2006.

[4] B. Altay and F. Başar, "Generalization of the sequence space $\ell(p)$ derived by weighted mean," *Journal of Mathematical Analysis and Applications*, vol. 330, no. 1, pp. 174–185, 2007.

[5] B. Altay and F. Başar, "On the paranormed Riesz sequence spaces of non-absolute type," *Southeast Asian Bulletin of Mathematics*, vol. 26, no. 5, pp. 701–715, 2003.

[6] V. Karakaya and H. Polat, "Some new paranormed sequence spaces defined by Euler and difference operators," *Acta Universitatis Szegediensis*, vol. 76, no. 1-2, pp. 87–100, 2010.

[7] V. Karakaya, A. K. Noman, and H. Polat, "On paranormed λ-sequence spaces of non-absolute type," *Mathematical and Computer Modelling*, vol. 54, no. 5-6, pp. 1473–1480, 2011.

[8] E. Malkowsky and E. Savas, "Matrix transformations between sequence spaces of generalized weighted means," *Applied Mathematics and Computation*, vol. 147, no. 2, pp. 333–345, 2004.

[9] B. Altay, F. Başar, and M. Mursaleen, "On the Euler sequence spaces which include the spaces ℓ_p and ℓ_∞. I," *Information Sciences*, vol. 176, no. 10, pp. 1450–1462, 2006.

[10] F. Başar and B. Altay, "On the space of sequences of p-bounded variation and related matrix mappings," *Ukrainian Mathematical Journal*, vol. 55, no. 1, pp. 136–147, 2003.

[11] M. Mursaleen and A. K. Noman, "On some new sequence spaces of non-absolute type related to the spaces ℓ_p and ℓ_∞ I," *Filomat*, vol. 25, no. 2, pp. 33–51, 2011.

[12] M. Mursaleen and A. K. Noman, "On some new sequence spaces of non-absolute type related to the spaces ℓ_p and ℓ_∞ II," *Mathematical Communications*, vol. 16, no. 2, pp. 383–398, 2011.

[13] M. Mursaleen, F. Başar, and B. Altay, "On the Euler sequence spaces which include the spaces ℓ_p and ℓ_∞. II," *Nonlinear Analysis*, vol. 65, no. 3, pp. 707–717, 2006.

[14] I. J. Maddox, "Spaces of strongly summable sequences," *The Quarterly Journal of Mathematics*, vol. 18, no. 2, pp. 345–355, 1967.

[15] I. J. Maddox, *Elements of Functional Analysis*, Cambridge University Press, Cambridge, UK, 2nd edition, 1988.

[16] I. J. Maddox, "Paranormed sequence spaces generated by infinite matrices," *Mathematical Proceedings of the Cambridge Philosophical Society*, vol. 64, pp. 335–340, 1968.

[17] K.-G. Grosse-Erdmann, "Matrix transformations between the sequence spaces of Maddox," *Journal of Mathematical Analysis and Applications*, vol. 180, no. 1, pp. 223–238, 1993.

Permissions

The contributors of this book come from diverse backgrounds, making this book a truly international effort. This book will bring forth new frontiers with its revolutionizing research information and detailed analysis of the nascent developments around the world.

We would like to thank all the contributing authors for lending their expertise to make the book truly unique.They have played a crucial role in the development of this book. Without their invaluable contributions this book wouldn't have been possible. They have made vital efforts to compile up to date information on the varied aspects of this subject to make this book a valuable addition to the collection of many professionals and students.

This book was conceptualized with the vision of imparting up-to-date information and advanced data in this field. To ensure the same, a matchless editorial board was set up. Every individual on the board went through rigorous rounds of assessment to prove their worth. After which they invested a large part of their time researching and compiling the most relevant data for our readers. Conferences and sessions were held from time to time between the editorial board and the contributing authors to present the data in the most comprehensible form. The editorial team has worked tirelessly to provide valuable and valid information to help people across the globe.

Every chapter published in this book has been scrutinized by our experts. Their significance has been extensively debated. The topics covered herein carry significant findings which will fuel the growth of the discipline. They may even be implemented as practical applications or may be referred to as a beginning point for another development. Chapters in this book were first published by Hindawi Publishing Corporation; hereby published with permission under the Creative Commons Attribution License or equivalent.

The editorial board has been involved in producing this book since its inception. They have spent rigorous hours researching and exploring the diverse topics which have resulted in the successful publishing of this book. They have passed on their knowledge of decades through this book. To expedite this challenging task, the publisher supported the team at every step. A small team of assistant editors was also appointed to further simplify the editing procedure and attain best results for the readers.

Our editorial team has been hand-picked from every corner of the world. Their multi-ethnicity adds dynamic inputs to the discussions which result in innovative outcomes. These outcomes are then further discussed with the researchers and contributors who give their valuable feedback and opinion regarding the same. The feedback is then collaborated with the researches and they are edited in a comprehensive manner to aid the understanding of the subject.

Apart from the editorial board, the designing team has also invested a significant amount of their time in understanding the subject and creating the most relevant covers. They scrutinized every image to scout for the most suitable representation of the subject and create an appropriate cover for the book.

The publishing team has been involved in this book since its early stages. They were actively engaged in every process, be it collecting the data, connecting with the contributors or procuring relevant information. The team has been an ardent support to the editorial, designing and production team. Their endless efforts to recruit the best for this project, has resulted in the accomplishment of this book. They are a veteran in the field of academics and their pool of knowledge is as vast as their experience in printing. Their expertise and guidance has proved useful at every step. Their uncompromising quality standards have made this book an exceptional effort. Their encouragement from time to time has been an inspiration for everyone.

The publisher and the editorial board hope that this book will prove to be a valuable piece of knowledge for researchers, students, practitioners and scholars across the globe.

List of Contributors

Ji-Huan He
National Engineering Laboratory for Modern Silk, College of Textile and Engineering, Soochow University, 199 Renai Road, Suzhou 215123, China

Hong Luo
College of Mathematics and Software Science, Sichuan Normal University, Sichuan, Chengdu 610066, China

Mohamed Jleli and Bessem Samet
Department of Mathematics, King Saud University, Riyadh 11451, Saudi Arabia

Erdal Karapınar
Department of Mathematics, Atılım University, 06836 ˙Incek, Ankara, Turkey

Hongjie Liu, Junqing Wang and Qiansheng Feng
School of Science, Tianjin Polytechnic University, Tianjin 300387, China

Zdeněk Šmarda
Department of Mathematics, Brno University of Technology, 61600 Brno, Czech Republic

Yasir Khan
Department of Mathematics, Zhejiang University, Hangzhou 310027, China

Vatan Karakaya
Department of Mathematical Engineering, Yildiz Technical University, Davutpasa Campus, Esenler 34220, Istanbul, Turkey

Necip Şimşek
Department of Mathematic, Istanbul Commerce University, U¨ sku¨dar, Istanbul, Turkey